T0133492

Basics of Polymer Chemistry

RIVER PUBLISHERS SERIES IN POLYMER SCIENCE

Indexing: All books published in this series are submitted to Thomson Reuters Book Citation Index (BkCI), CrossRef and to Google Scholar.

The "River Publishers Series in Polymer Science" is a series of comprehensive academic and professional books which focus on theory and applications of Polymer Science. Polymer Science, or Macromolecular Science, is a subfield of materials science concerned with polymers, primarily synthetic polymers such as plastics and elastomers. The field of polymer science includes researchers in multiple disciplines including chemistry, physics, and engineering.

Books published in the series include research monographs, edited volumes, handbooks and textbooks. The books provide professionals, researchers, educators, and advanced students in the field with an invaluable insight into the latest research and developments.

Topics covered in the series include, but are by no means restricted to the following:

- Macromolecular Science
- Polymer Chemistry
- Polymer Physics
- Polymer Characterization

For a list of other books in this series, visit www.riverpublishers.com

Basics of Polymer Chemistry

Muralisrinivasan Natamai Subramanian

Consultant, India

Published, sold and distributed by:
River Publishers
Alsbjergvej 10
9260 Gistrup
Denmark

River Publishers
Lange Geer 44
2611 PW Delft
The Netherlands

Tel.: +45369953197
www.riverpublishers.com

ISBN: 978-87-93519-01-5 (Hardback)
978-87-93519-02-2 (Ebook)

Contents

Preface

Polymer chemistry clearly has become a broader discipline. It has had a greater influence on civilization than any other technological discipline. New polymers arise from new discoveries. Basics of polymer chemistry covers its importance and give an overview but is certainly deeper insight into the chemistry. All major polymers covered, with their chemical structures and some aspects of synthesis, but most space is devoted to mainly descriptive and many practical hints and useful information. All chapters with an extensive list of references, which is very useful. Basics of polymer chemistry is comprehensive which represent a growing a field of multidisciplinary research ranging synthesis to manufacture. The objective of the book is to provide basic research at universities, research institutes and industrial laboratories. The basic chemistry research are oriented toward the marketplace and to the needs of the humankind.

Basics of polymer chemistry for polymer courses develop a toolbox of polymer knowledge for the study and workplace. This book reinforces industrially important polymer chemistry and provides an intuition for polymer structure relate to material properties. The academic and research topics in polymer chemistry range from structure to large-scale engineering. The achievements of polymer in different fields are evident all around us.

This book covers most of the chemistry related to University syllabus and can be extremely useful for studies in polymer and for experienced users. The potential subject matter is wide that coverage is going to be superficial and intended as admirable contents for both chemists and engineers.

This increased coverage is directed towards discussing the basic approach is described related to the chemistry. This book is developed for polymer chemistry provides the following six dimensions:

- Polymer chemistry as aspects of chemistry
- Basics of polymer, structure and properties
- Basics of polymerization processes

- Basics of chemical reactions involved in the polymerization
- Technological manifestations of polymer chemistry
- Basis for the implementation to be as text or reference materials.

This book can serve as the eductional material. This book is an effective outcomes as cognitive measures. Increasing emphasis has been placed on the necessity to match the design of polymer chemistry curricula. This book have an impact material on research in polymer chemistry, and also in curricula and instructional material. This book also as a curricula in order to:

- Show the application of basic chemical principles in polymer production
- Develop a basic knowledge of the importance of the technological and economic factors
- Demonstrate the uses of polymer to society.

Polymer chemistry is a very wide subject of polymers reactions. Polymer syntheses are in a separate section covering only those aspects of chemistry.

Dr. Muralisrinivasan Natamai Subramanian
Madurai

Acknowledgement

I would like to acknowledge the following for their support during the writing of this manuscript:

- Mrs. Himachala Ganga, Mr. Venkatasubramanian and Mr. Sailesh is encouraged me in every steps of writing book.
- My teachers who taught me in college education provided the encouragement to get the job done and help to bring this book.
- Thanks to Mr. Mark for giving me the opportunity to publish this book originally and encouraging me also.
- Additional thanks to Mrs. Junko, who involved in bring this book to College and University education.
- Overall thanks go to my parent and God who gave their birth in the earth.
- Above all, I pray to the almighty Nataraja to bring me in the wonderful earth to complete the jobs.

List of Figures

List of Tables

Abstract

The pioneers of polymer chemistry realized that the chemistry of polymer containing compounds offers multiple advantages. Monomers can be polymerized using polymerization techniques. It is interesting to analyse that not all monomers respond well to each types of polymerization techniques. Polymer is commonly a large molecule composed of repeating units connected in a variety of ways. Simple linear polymer is the repeating unit connected in a linear sequence. An alternative to a linear polymer is a branched one. The branches can be long or short. Repeating unit of polymer depends on the structure of the starting monomer, the initiating system, and the condition of polymerization reactions.

A wide range of reactions is possible, and can be carried with catalysts and initiators. The various findings in, and applications of, polymer chemistry has led, in recent years, to a tremendous growth which is reflected by the large number of publications in this field. The physical properties of a polymer are based on the configuration of the constituent atoms, and to some extent by the molecular weight. The configuration is partly dependent on the main chain, and partly on the various side groups.

Most of the polymers are based on long chains of carbon atoms. High cohesive energy associates with the chain leads to high melting point and may be associated with crystallinity and low cohesive energy chains in the polymer have low softening point and easy deformation by applied stresses. The properties of polymers are governed to some extent by molecular weight as well as molecular structure. Properties also depend partly on the distribution of molecular weights, and in copolymers on the distribution of molecular species. By emphasizing the polymerization and polymer synthesis, this book presents a comprehensive overview of the various facets, and scopes of polymer chemistry.

This book is a comprehensive description and provides up-to-date-information concerning the monomers, polymers, synthesis, and reactions. It will show the importance of the polymer chemistry to the industry.

1

Introduction

Polymers are macromolecules with accurately controlled structure. The well-defined structure is an important aspect of polymer chemistry. Today to the society and industry, polymers are essential to life for comfort, hygiene, and well-being. Polymers are necessary along with high-value added applications in our life. The applications are different from high-tech science to consumer products with economy.

Most of the fields in science and engineering are connected in some way with polymers. Yet only the field of polymer chemistry focuses directly on them. These broad disciplined studies related to atoms about how it can be assembled into the polymer structure and how these structures and properties can be engineered to useful ends.

Polymer chemistry has assumed a growing relevance, particularly in the last two decades, as a valid tool in the synthesis or modification of molecules suitable for different applications. In fact, polymer chemistry comes into use in different areas. The number of applications as well as potential interest of polymer chemistry lies in the versatility of their synthesis from chemicals and on the wide range of reactions possible.

Polymer chemistry related to large-scale achievements in the field is evident all around the industry. Polymers are almost entirely the result of material improvements such as biomaterials from the basis for medical implants and artificial organs. They are revolutionizing drug delivery and cancer therapy and decreasing the environmental impact of manufacturing.

Polymer improvements are evident from the electronics industry to the automobile industry and also to the medical applications. Before the invention of polymers, natural materials such as wood, wool, silk and products such as glass, leather, metals, and alloys are for ingenuity. Natural polymers are originally from plants, animals, etc. but because of their biodegradable nature, their uses are limited in number.

1.1 History

In 1939, Simon observed the change of liquid styrene which changed into solid upon being heated. Bayer recorded the reactions of phenols and aldehydes resulting in the formation of resinous substances [1, 2]. During that period people did not have any idea about the polymer.

In the 1920s, the concept of macromolecules proposed by Staudinger. Carothers succeeded in the synthesis of polyesters in 1930s and worked out the theoretical models for condensation polymerization. In modern days, polymers are categorized as natural as well as synthetic in nature. In synthetic polymers, the synthesis and manufacturing of polymers are to accelerate the development of new applications with value addition [3, 4]. A synthetic polymer's discovery was purely accidental due to the complexity of the manufacturing process. Polymerization reactions can be traced to as early as the nineteenth century.

Historically, polymers only contained the elements C, H, N, O, S, Cl, Br, and occasionally P. The rapid development of polymer materials is expanded further the list to silicon as an inorganic element. However the vast majority of elements in periodic table have not found their way into polymers. The introduction of metals in polymer chemistry particularly into polymerization through the use of co-ordination chemistry. Metals are introduced in polymer by metal chelation reactions. However, many of the times, useful materials did not emerge from the reactions due to intractable low molecular weight material with not possible to characterize along with insolubility which is a major problem [5, 6].

1.2 Organic Molecules – Polymer Basis

Organic molecules contain carbon covalently bonded to other atoms that determine the structure and function. Inorganic molecules do not contain carbon and hydrogen together they do have other important roles (water, salts, and many acids and bases). In organic chemistry, carbon is the living element which is electro neutral that never loses or gains electrons, but shares with other elements.

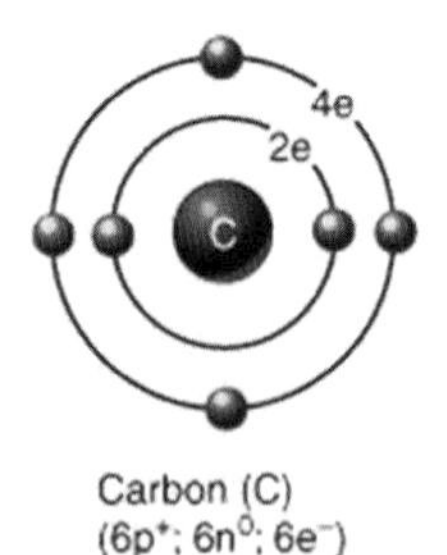

Thereby, the carbon atom shares its electron with the atoms of different elements to form a polymer chain with its 4 electrons, and requires 8 more electrons to fill its valence shell. It can form strong, stable covalent bonds with 4 other elements usually H, O, N, S, P or with another C. The uniqueness of Carbon is that it can bind to itself and can form long chains. Many different Carbon-based polymers have unique structures.

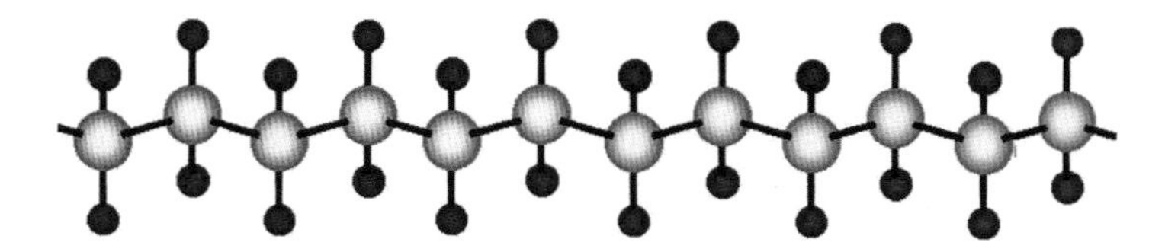

Nowadays, the polymers offer a great technological application potential in several areas [7–9] such as pharmaceutics apparatus, protection against corrosion, building and construction, packaging, automobiles, defence products, etc. Polymer chemistry of stable polymers contains covalent or coordination bonds between the monomers. Polymers have electron deficient bonds (borates, polymeric palladium chloride) or hydrogen bonds (polymeric water, liquid hydrogen fluoride, crystallized methanol) are also considered as polymers. However, the monomers are connected with the covalent bonds in polymer.

Polymers have developed as a part of macromolecules providing the chemistry with tools and at the same time utilizing synthesis, polymerization, and application which cause a variety in development. New polymerization techniques are developed to identify the applications in various areas ranging from consumer to engineering to space. Thus, polymer chemistry remains in the forefront of science.

1.3 Summary

- Polymers are macromolecules with distributed composition and chain lengths.
- Polymers are produced in a broad spectrum which is indispensably used in all parts of life.
- The demand on polymers has increased which causes tremendous pressure to make production more efficient along with cost effective methods.
- Polymer products are enabled in compliance with the product characteristics based on market demand and economy.
- Increased polymers help to recycle the same by saving all the renewable resources.

References

[1] E. Simon. *Ann. Chim. Phys.*, **31**, 265 (1839).
[2] K. J. Saunders. *Organic Polymer Chemistry: An Introduction to Organic Chemistry of Adhesives, Paints, Plastics, and Rubbers*, Chaman and Hall: London (1973).
[3] H. Staudinger. *Ber. Dtsch. Chem. Ges.*, **53**, 1073 (1920).
[4] W. H. Carothers. *Chem. Rev.*, **8**, 353 (1931).
[5] E. Rochow, *Chem. Eng. News*, **23**, 612 (1945).
[6] F. G. A Stone and W. A. G. Graham. *Inorganic Polymers*, Academic Press, New York (1962).
[7] P. F. W. Simon, R. Ulrich, H. W. Spiess, U. Wiesner. *Cem. Mater*, **13**, 3464 (2001).
[8] M. R. Saboktakin, A. Maharramov, M. A. Ramazanov. *Nat. Sci.*, **5**, 67–71 (2007).
[9] M. R. Saboktakin, A. Maharramov, M. A. Ramazanov. *J. Am. Sci.*, **3**, 40–45 (2007).

2

Monomers

Monomers are molecules, which are the primary materials that react selectively under polymerization reaction. Molecules which are suitable for the formation of macromolecules are termed as monomers. Monomers are relatively small molecules. These monomers can have the desired glass transition temperature, flexibility, mechanical strength, polarity and hydrophilic or hydrophobic character of the resulting polymer.

Monomers can be easily polymerized into polymers. Monomers are the primary materials in the polymerization process to form a polymer. They are the building blocks of polymers, and they get formed in the reaction. They are chemical molecules [1]. Monomers join with a covalent bond to form a polymer. Constitution and configuration of monomers get reflected in the mechanical properties of polymers. These monomers are versatile in the group whose chemical properties are primarily determined by the nature of the polymer.

2.1 Classification of Monomers

Monomers classify as:

1. Monomers with unsaturation molecule undergo chain growth polymerization – molecules contain double or triple bonds such as ethylene, propylene, styrene, etc.
2. Mono or bifunctional molecules may undergo polymerization reaction by step-growth polymerization with either elimination or without elimination reaction – mono or bifunctional groups of single or different molecules of reactive groups such as acids, amines, and hydroxyl group.
3. Monomers with a ring opening – undergo reaction to form polymers such as caprolactam, etc.,

Further classification is based on the activity of the monomer [2].

1. More activated monomers include vinyl monomers (styrene and vinylpyridine), methyl methacrylate (MMA), methacrylic acid, and methacrylamide, and acrylics (methacrylate, acrylic acid, acrylamide and acrylonitrile.
2. Less activated monomers such as vinyl esters such as vinyl acetate and vinyl amide such as *N*-vinylpyrrolidone (NVP) and *N*-vinylcarbazole (NVC).

2.2 Monomers and Their Reactions

Monomers such as acetylene, ethylene, propylene, etc., without any functional group undergo reactions with a cleavage of their double or triple bonds. The monomers of linear molecules undergo polymerization yields a linear polymer. Linear molecules with bifunctional molecules undergo reactions within themselves such as with hydroxyl acid or with other bifunctional molecules which form branched or cross-linked polymers. Monomers contain more than one secondary reactive group such bi, tri or poly functional groups which can be elaborated in multiple ways to form branched or cross-linked polymers during polymerization. Chemical reactions between monomers of mono functional or bifunctional groups get induced by several means to form polymers. Monomer possesses functional groups reacting selectively under the selected conditions.

2.2.1 Unsaturated Monomers

Hydrocarbons contain only carbon (C) and hydrogen (H). Hydrogen atom in hydrocarbons gets replaced by other atoms or groups of atoms, and functional groups. They are the reactive sites in molecules. The majority double and triple bonds in monomers are considered as a reactive center in polymerization reactions. The carbon–carbon double bond is important in polymer synthesis and manufacturing, not only for the obvious reason that the compound is utilized but also for the formation of the double bond which allows the introduction of a wide variety of functional groups to undergo polymerization reactions.

2.2.2 Monomers with Hydroxyl Groups

Hydroxyl groups have utility as hydrogen bonding sites and can provide polymers with compatibility for water or polar solvents. These versatile functional groups can broadly form polymer. Polymers containing free –OH groups can be post reacted with acids, epoxies, isocyanates, etc. to create novel polymer properties and architectures.

2.2.3 Monomers with Acidic Groups

Acidic groups have been utilized to a wide range of alternative functional groups. They function as catalysts for chemical reactions. The acidic groups are employed as a function group. In anhydrides and esters, the acid group is protected and hence it is not useful as a cross-linker.

2.2.4 Olefinic Monomers

Monomers consist only of carbon and hydrogen. They are available freely from petroleum resources as feedstock. They are gaseous in nature. From the petroleum feed stocks, ethylene, and propylene are to be separated and utilized as monomers to produce polymer. Materials made from olefinic monomers have typical chemical resistance and are very hydrophobic. Some of the monomers are

$H_2C{=}CH_2$	$CH_3{-}CH{=}CH_2$	$H_2C{=}CH{-}CH{=}CH_2$
Ethylene	Propylene	1,3-butadiene

Highly substituted olefinic molecules are more stable.

$$\text{tetra} > \text{tri} > \text{di} > \text{mono}$$

2.3 Amine

Amines are most widely used as monomers with versatile functional groups. Di-functional monomers are useful for imparting crosslinking or branching to polymer architecture. Cross-linked polymer structure results from the spacer group that helps to determine the physical and mechanical attributes. Amino monomer such as amine or amide with an aldehyde undergoes polycondensation or by curing reactive oligomers prepared from an amino component and an aldehyde.

2.3.1 Aniline

In the chemistry of aniline, the synthesis and chemical properties of polymers are present [3] and in the treatment of hardening, aniline with formaldehyde it forms the network structure [4]. The synthesis of urea resin and related products use an amidomethylation similar to phenol formaldehyde such as

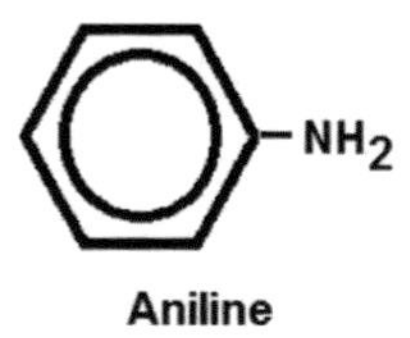

Aniline

hardening of phenol formaldehyde by hexamethylenetetramine [5] involves both aminomethylation and substitution reactions as well as a catalytic effect. Aromatic amine monomers are more reactive in polymerization reactions.

2.3.2 Acrylamide and N-substituted Acrylamides

Acrylamide and N-substituted acrylamides used in both fine organic synthesis and industry as intermediate products for the preparation of carbo- and hetero-chain polymers and copolymers and polyfunctional acyclic and heterocyclic compounds. Acrylamide is a chemical compound that is unique among vinyl monomers. It has a chemical formula of C_3H_5NO. It is a white odorless crystalline solid. It is soluble in water, ethanol, and chloroform. Acrylamide polymerized to ultrahigh molecular weight [6]. It gets produced from industrial processes. It is neurotoxic as a possible carcinogen in humans.

$$H_2C{=}CH\overset{\overset{O}{\|}}{C}{-}NH_2$$

Acrylamide

2.4 Butyl Acrylate

Butyl acrylate is a colorless liquid. It has a fruity odor and miscible readily with most organic solvents. The structure of butyl acrylate is $CH_2{=}CHCOOC_4H_9$. It has a boiling point of 147°C. Based on the monomer and reaction conditions, it polymerizes typically to form acrylic polymers or terpolymers with wide range of properties. It has wide range of application as industrial chemicals [7]. It polymerizes readily and displays a wide range of properties dependent upon the selection of the monomer and reaction conditions. It forms homopolymers and copolymers. Copolymers of butyl acrylate can be prepared with acrylic acid and its salts, amides, and esters, and with methacrylates, acrylonitrile, maleic acid esters, vinyl acetate, vinyl chloride, vinylidene chloride, styrene, butadiene, and unsaturated polyesters. It is a useful feedstock for chemical syntheses, because it undergoes addition reactions with a variety of organic and inorganic compounds. The structural formula is as follows:

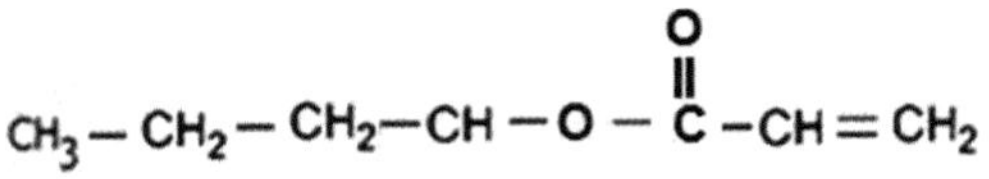

$$CH_3-CH_2-CH_2-CH-O-\overset{\overset{\large O}{\|}}{C}-CH=CH_2$$

Butyl acrylate

2.5 Butadiene

Butadiene is conjugated as diene of structure $H_2C = CH - CH = CH_2$. It has geometrical isomers of cis and trans form. It is an important constituent of many synthetic polymers. It is a colorless gas. It can be liquefied either by cooling to –4.4°C or by compressing to 2.8 times at ambient temperature. It replaces natural rubber.

$$H_2C=CH-CH=CH_2$$

Butadiene

Natural gas is a major raw material and to some extent forms ethyl alcohol. Almost butadiene made by dehydrogenation of butane or butylene or high pressure cracking of petroleum distillates. It has a planar structure with low resonance energy [8–11]. Butadiene undergoes emulsion polymerization into poly(butadiene).

2.6 Divinylbenzene

Divinylbenzene is a commercial mixture of m- and p-divinylbenzene, ethylvinylbenzenes, and diethylbenzenes. It is produced by hydrogenation of an isomeric mixture of diethylbenzenes. It is used as a crosslinking agent in large number of different polymeric materials.

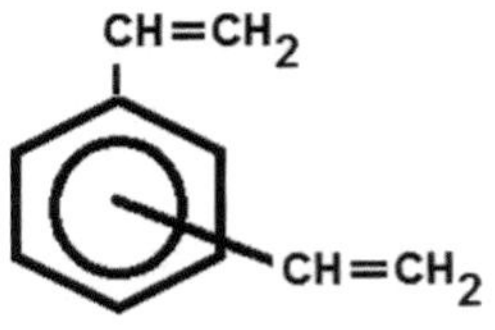

$CH=CH_2$

$CH=CH_2$

Divinylbenzene

Thermal polymerization produces polymer which are brittle and highly cross-linked polymer [12]. Under specific conditions, it forms soluble homopolymers of divinylbenzenes by radical, cationic or anionic mechanisms.

2.7 Esters of Acrylic and Methacrylic Acid

Esters of acrylic acid are monomers readily combine with themselves or with other monomers to form long chains of repeating units or polymers. The esters of acrylic and methacrylic acid monomers are unsymmetrically substituted ethylenes of the general formula

$$H_2C{=}C(R)OC(=O)R'$$

With R=H for acrylates and R=CH3 for methacrylates. The substituents R' may be of a great variety from n-alkyl chains to more complicated functional groups. These compounds are generally named as acrylic esters, where R is an alkyl group. Acrylic and methacrylic acid (propenoic and 2-methylpropenoic acid) are the basic compounds of a large number of derivatives, such as esters, amides, nitriles, chlorides and aldehydes are also being used as monomers to synthesis polymer [13–16]. The derivative of acrylic acid, methacrylic acid, and related monomers form the acrylic polymer. Some of the polymers of esters of acrylic acid and methacrylic acid are given below:

$$\left[-C(H)(COOH)-CH_2-\right]_n$$

poly(acrylic acid)

$$\left[-C(CH_3)(COOCH_3)-CH_2-\right]_n$$

poly(methyl methacrylate)

$$\left[-C(H)(CONH_2)-CH_2-\right]_n$$

polyacrylamide

$$\left[-C(H)(CN)-CH_2-\right]_n$$

polyacrylonitrile

$$\left[-C(CN)(COOC_2H_5)-CH_2-\right]_n$$

poly(ethyl cyanoacrylate)

2.8 Methacrylic Ester Monomers

Methacrylic ester monomers are a versatile group of monomers whose chemical properties are primarily determined by the nature of the R side chain group in the structure:

$$H_2C{=}C(CH_3)OOR$$

The chemical properties, molecular weight and tacticity governed by the presence of – R side chain group in the structure. Methyl methacrylate undergoes bulk polymerization and forms poly(methyl methacrylate).

$$H_2C{=}CCH_3\,O\overset{\overset{\displaystyle O}{\|}}{C}CH_3$$

Methyl methacrylate

Methacrylic ester with R $-CH_3$ in the position of the vinyl group leads stability, hardness, and stiffness to the polymer. A polymer's physical and chemical properties also depend upon the R group, the molecular weight, and the tacticity of the polymer. Methacrylics will readily polymerize with other methacrylic and acrylic monomers. This ability to form copolymers allows polymers to be created as polymers ranging from tacky adhesives to hard powders and rigid sheets.

2.9 Ethylene

Ethylene has a chemical structure of $CH_2 = CH_2$. Ethylene is a gas from natural sources includes natural gas and petroleum. It is an important organic chemical useful in the conversion of polyethylene by high pressure or low-pressure process. Ethylene is produced commercially by the steam cracking of hydrocarbon feedstock. However, ethylene mainly obtained from naphtha cracking, condensates with coproduction of propylene, and pyrolysis of gasoline. The melting point of ethylene is –169.4°C and its boiling point is –103.9°C [17]. Ethylene undergoes coordination polymerization.

Ethylene monomers are of great interest due to reduced melting point and crystallinity of a polymer at low concentrations. Longer-chained monomers such as 1-hexene are more effective. It results in a branched polyethylene with methyl branching if propylene is used, ethyl branching if butylene is used, and so on.

Ethylene (C_2H_4) contains a carbon–carbon double bond in its most stable Lewis structure, and each carbon has a completed octet.

$$H_2C=CH_2$$

Ethylene

2.10 Ethylene Glycol

Ethylene glycol is the simplest diol with a clear, colorless, odorless liquid with a sweet taste. It is hygroscopic and completely miscible with many polar solvents such as water, alcohols, glycol ethers, and acetone. Its solubility is low, however, in nonpolar solvents, such as benzene, toluene, dichloroethane and chloroform. Ethylene glycol can be alkylated or acylated to form ethers or esters respectively. The presence of two hydroxyl groups leads to the formation of both mono and diethers and mono and diesters. The esterification of ethylene glycol with terephthalic acid to form polyesters by the catalyst system that is used and glycol excess.

$$HOCH_2-CH_2OH$$

Ethylene glycol

Ethylene glycol units are strongly hydrophilic through their multiple H-bonding sites. Monomers of this type are useful in the construction of hydrogels and water compatible polymer structures. Bioactive molecules e.g. drugs with attached polyethylene glycol (PEG) chains have improved bioavailablity characteristics. Vinyl and ethenyl monomers are often used to create polymers with inert main chain features. Many of these monomers polymerize through metellocene or other metal mediated polymerization processes.

2.11 Maleic Anhydride (MA)

Maleic anhydride (MA) has been widely used as a grafting monomer. It is to functionalize polyolefins because of the higher reactivity of the anhydride group [18–20].

Maleic anhydride

2.12 *N*-Vinylcarbazole (NVC)

NVC is a white crystalline material that melts at 65°C. It is soluble in the most aromatic, chlorinated, and polar organic solvents. The material is handled with care because it may cause severe skin irritations. Like other aromatic amines, NVC is suspected to cause cancer [21]. Most published synthetic routes use carbazole, which is readily available from coal-tar distillation, as the starting material. The industrial process is the vinylation reaction of carbazole and alkali metal hydroxide or preformed alkali metal salts of carbazole with acetylene [22, 23]. The reaction is normally carried out at high boiling points.

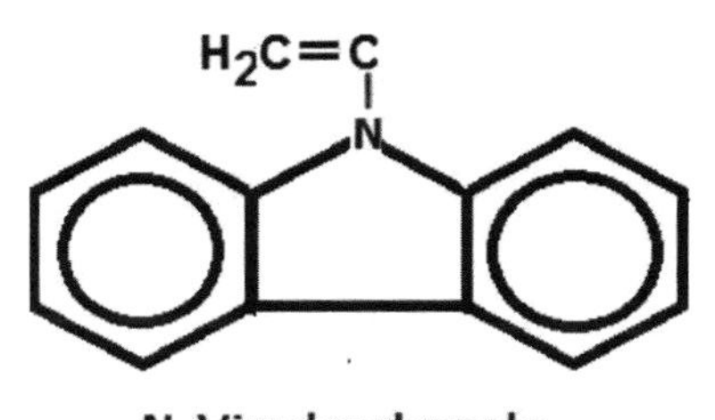

N-Vinylcarbazole

2.13 *N*-Vinylpyrrolidone (NVP)

NVP is one of the various products of the acetylene chemistry [24] discovered by Reppe [25–28]. The reaction between acetylene and formaldehyde yields 1,4-butinediol, which is hydrogenated to 1,4-butanediol. After the oxidative cyclization to γ-butyrolactone on Cu contact (700°C), the reaction with ammonia leads to γ-hydroxycarbonamide, solvents at slight pressures.

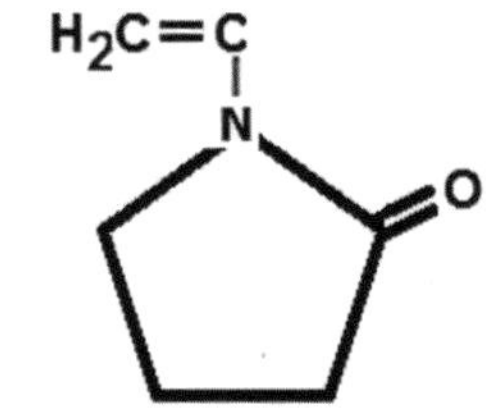

$H_2C=C$

N

O

N-Vinylpyrrolidone

2.14 Propylene

Stereoisomerism is the structural variation in the position of small side groups along with the long chain polymer arising from the presence of asymmetric centers formed during polymerization. Substituted methyl group in ethylene is termed as propylene monomer, which becomes asymmetric carbon in a polypropylene that gives tacticity to the polymer. Propylene undergoes coordination polymerization in order to stereo specific polymers.

$$H_2C=CH-CH_3$$

Polypropylene

Polypropylene (PP) is a widely used polymer because of its outstanding mechanical properties and low cost [28]. However, PP has the disadvantage of becoming brittle at low temperatures because of its high transition temperature and high crystallinity. Polymerization of propylene monomer undergoes co-ordination polymerization [29].

2.15 Styrene

Styrene is an aromatic hydrocarbon [30]. Styrene undergoes bulk and suspension polymerization to produce polymer. It also undergoes cationic and anionic polymerization reactions.

$H_2C=CH$

Styrene

Styrenic monomers generally provide polymers of higher glass transition temperature, higher modulus, increased hydrophobic character and nominal UV absorbance. Cross-linked styrene resins (especially in microsphere form) are tough and chemically resistant. These form the basis for ion exchange resins and microbeads used as supports for biochemical reactions.

2.16 α-Methylstyrene

α-methylstyrene is a derivative of styrene. It is produced commercially by the dehydrogenation of cumene and as a by-product in the production of phenol and acetone by the cumene oxidation process. The polymerization characteristics of α-methylstyrene are different from those of styrene.

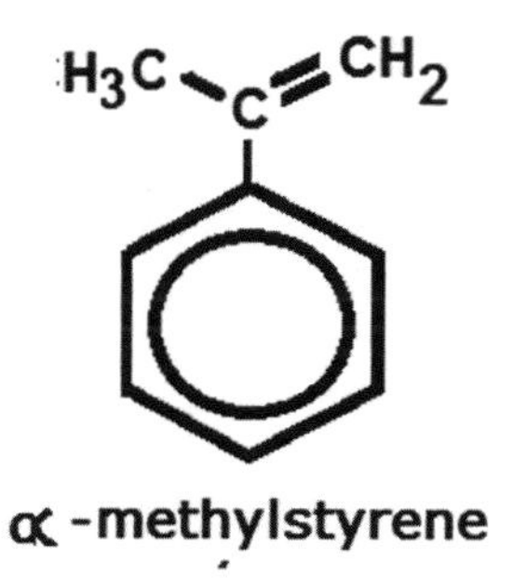

Radical polymerization of the pure monomer proceeds very slowly and is not a practical technique, both ionic and coordination type polymerization can be used to prepare poly(α-methylstyrene) [31].

2.17 Vinyl Arenes

Vinyl arene monomers generally prepared by dehydration of the corresponding carbinol, which usually obtained by acetylation of the corresponding arene and reduction of the ketone. The carbinol obtained by the reaction of the arene Grignard reagent with a carbonyl compound.

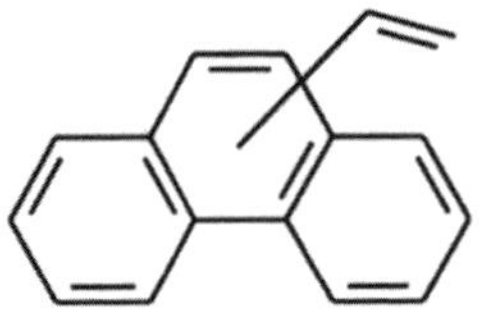

vinylphenanthrene 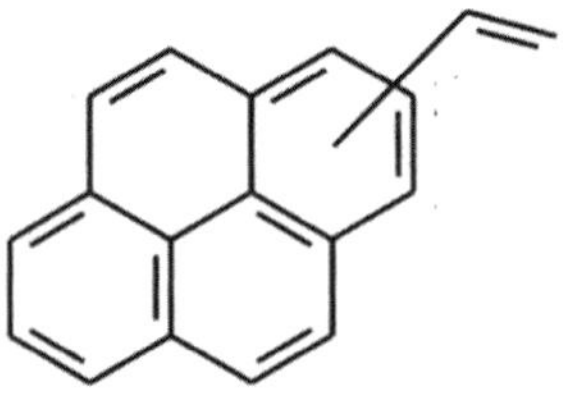 vinylpyrene 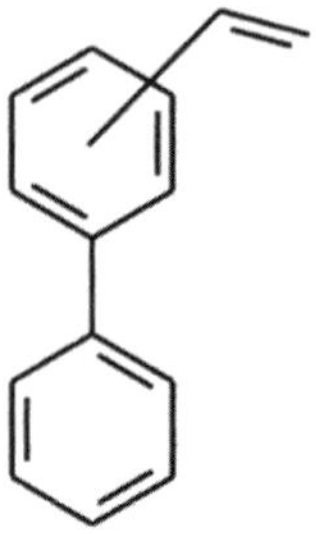 vinylbiphenyl

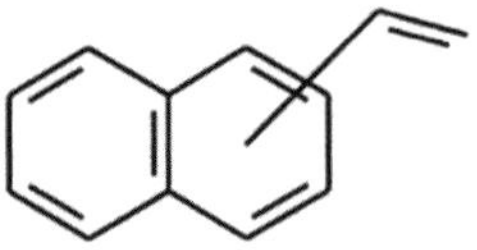

vinylnaphthalene 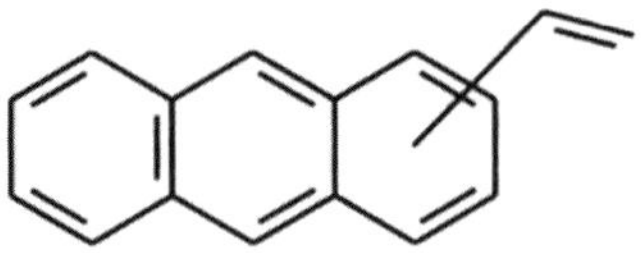 vinylanthracene

2.18 Vinyl Chloride

The structure of vinyl chloride is $CH_2 = CHCl$. It has boiling point of –13.9°C. Therefore, it has high vapor pressure at ambient temperature. Vinyl chloride undergoes reactions by suspension or emulsion techniques in the manufacture of polyvinylchloride. Vinyl chloride is a colorless gas and useful in the production of polyvinylchloride.

$$CH_2{=}\ CHCl$$

Vinylchloride

The catalytic hydrogenation of acetylene to vinyl chloride is not economical due to higher energy costs. There are other methods of manufacturing vinyl chloride through direct chlorination or oxychlorination. Initial formation of ethylene dichloride is subjected to thermal cracking to form vinyl chloride [32].

2.19 Vinyl Ethers

Vinyl ethers comprise that class of olefinic monomers, which possess a double bond situated adjacent to ether oxygen. These monomers include those compounds, which have various substituents attached to the carbon atoms of the double bond, as well as the unsubstituted compounds. Due to the presence of the neighboring oxygen atom, the double bond possesses a highly electronegative character, a feature that dominates both the organic and polymer chemistry of these compounds. The analogous vinyl thioethers are also known [33] and their chemistry closely parallels that of their corresponding vinyl ether counterparts.

$$RO-CH=CH_2$$

Vinyl ethers

The oldest, most versatile and major commercial method for the synthesis of vinyl ethers is by the base-catalyzed condensation of acetylene with alcohols first described by Reppe and co-workers [34, 35]. Ethyl vinyl ether undergoes violent exothermic polymerization with elemental iodine that catalyses the reaction [36]. The polymers produced from ethyl vinyl ether monomer by cationic polymerization are versatile. The polymer has wide variety of technical applications [37–42].

2.20 Network-Forming Monomers

2.20.1 Epoxides

The reactive epoxides are reaction product of epichlorohydrin with a di or multifunctional acid, alcohol, amine or phenol, most often with 4,4' – (propane – 2,2' diyl) diphenol(bisphenol A).

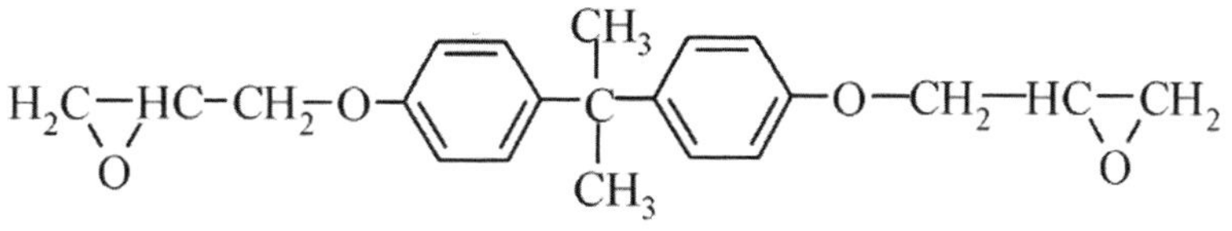

$H_2C-HC-CH_2-O-C_6H_4-C(CH_3)_2-C_6H_4-O-CH_2-HC-CH_2$ (each terminal H_2C-HC bridged by O as oxirane ring)

2,2' – bis[4 -(2,3 – epoxypropoxy)phenyl]propane

2,2'- [Propane 2,2-diyl)bis(4,1 – Phenylenenoxymethylene)bis(oxirane)

2.21 Summary

1. Monomers are the most fundamental molecules of synthesizing polymers.
2. Economic importance and availability of high-volume petrochemical monomers, such as ethylene, propylene, and other higher α-olefins, as

well as emerging discoveries in olefin polymerization processes and catalysis.

3. The preparation conditions in terms of temperature and time, as well as starting monomers control the structure and properties of the polymer.
4. Polymer characteristics are based on primary monomers. The characteristic features of the chemical constitution of molecules have gained general acceptance in polymer chemistry.
5. Based on chemical structures, the monomers are acrylic, diene, phenolic, and vinylic. The backbones of the polymer molecules such as amide, ester, ether and ring structures such as benzimidazole, benzoxazole, quinoxaline are the chemical characteristic groups. Linear, branched, dendritic and comb are overall shapes of the polymer molecules based on the molecular architecture.

References

[1] G. Gruenwald. *Plastics, how structure determines properties.* Hanser, Munich (1992).
[2] Graeme Moad, Ezio Rizzardo, San H. Tang. *Acc. Chem Res.*, **41**, 9, (2008) 1133–1142.
[3] K. J. Saunders. *'Organic Polymer Chemistry'*. Chapman and Hall, London, 316 (1977).
[4] K. J. Saunders. *'Organic Polymer Chemistry'*. Chapman and Hall, London, 375 (1977).
[5] K. J. Saunders. *'Organic Polymer Chemistry'*. Chapman and Hall, London (1977), 286–295.
[6] W. M. Kulicke, R. Kniewske, J. Klein. *Prog. Poly. Sci.*, **8**, 873 (1982).
[7] K. W. Min, W. H. Ray. *J. Macromol. Sci., Rev. Macromol. Chem.*, **11**, 177 (1974).
[8] J. W. Mcbain, J. J. O'Connor. *J. Am. Chem. Soc.*, **63**, 875 (1941).
[9] D. J. Marais, N. Sheppard, B. P. Stoicheff. *Tetrahedron*, **17**, 163 (1962).
[10] J. J. Fisher, J. Michl. *J. Am. Chem. Soc.*, **109**, 1056 (1987).
[11] K. B. Wiberg, R. E. Rosenberg, P. R. Rablen. *J. Am. Chem. Soc.*, **113**, 2890 (1991).
[12] K. E. Coulter, H. Kehde, and B. F. Hiscock. *Vinyl and Diene Monomers, Part 2* (Leonard, E. C., ed.). Wiley-Interscience, New York, 540 (1971).
[13] H. V. Pechmann, O. Röhm. *Ber. Dtsch. Chem. Ges.*, **34**, 427 (1901).
[14] Röhm, and D. E. Haas. 571, 123, US 1 829 208 (W. Bauer). (1928).

[15] L. S. Luskin. *"Vinyl and Diene Monomers," in C. Leonhard (ed.): High Polymers, Part I*, Wiley Interscience, New York, **24**, 105 (1970).
[16] H. Spoor. *Angew. Makromol. Chem.*, **4/5**, 142 (1968); H. Wesslau. *Makromol. Chem.*, **69**, 220 (1963).
[17] H. K. Stryker, G. J. Mantell, A. F. Helin. *J. Appl. Polym. Sci.*, **11**, 1 (1967).
[18] N. G. Gaylord, M. Mehta. *J. Polym. Sci. Plym. Lett. Ed.*, **20**, 481 (1982).
[19] N. G. Gaylord, M. Mehta, R. J. Mehta. *Appl. Polym. Sci.,* **33**, 2549 (1987).
[20] Y. Minoura, M. Ueda, S. Mizunuma, M. Oba. *J. Appl. Polym. Sci.,* **13**, 1625 (1969).
[21] H. J. Klimisch, and H. Kieczka. *Ullmanns Enzyklopädie der technischen Chemie* (Bartholomé, E., Biekert, E., Hellmann, H., Lei, H., Weigert, W. M., and Wiese, E., eds.), Verlag Chemie, Weinheim, 4th ed., **23**, 597 (1983).
[22] W. Reppe, and E. Keyssner. *Ger. 618,120 to IG Farbenindustrie* (1935); C.A., **30**, 110 (1936).
[23] W. Reppe, and coworkers. *Justus Liebigs Ann. Chem.*, **601**, 81 (1956); H. Davidge. *J. Appl. Chem.*, **9**, 241 (1959).
[24] L. N. Bauer, R. B. Healy, and H. R. Stringer. *Ger. to Röhm & Haas Co.* (1959); C.A., **55**, 14997i (1961).
[25] F. Haaf, A. Scanner, and F. Straub. *Polym. J.*, **17**, 143 (1985).
[26] W. Reppe. *Polyvinylpyrrolidon*, Verlag Chemie, Weinheim (1954); see. *Angew. Chem.*, **65**, 577 (1953).
[27] W. Reppe, and coworkers. *Justus Liebigs Ann. Chem.*, **601**, 81, 135, 481 (1956).
[28] C. Harper. *Handbook of Plastics and Elastomers*. McGraw-Hill Inc., New York (1975).
[29] N. G. Gaylord, H. F. Mark, N. M. Bikales. *Encyclopedia of polymer science and technology*. Wiley Interscience, New York, **11**, 597 (1970).
[30] F. Bovey, I. Kolthoff. *J. Polym. Sci.*, **5**, 487 (1950).
[31] K. E. Coulter, and H. Kehde. *Encyclopedia of Polymer Science and Technology*, (Mark, H. F., Gaylord, N. G., and Bikales, N., eds.), Wiley-Interscience, New York, **13**, 151 (1970).
[32] E. A. Grulke. *In Kirk-Othmer Encyclopedia of Chemical Technology* (Mark, H., ed.), Interscience Publishers, New York, **1**, 225 (1963).
[33] C. E. Schildknecht. *Vinyl and Related Polymers*. Wiley, New York, 593 (1952).

[34] W. Reppe. U.S. 1,959,927 (1934); C.A. **28**, 4431 (1935).
[35] W. Reppe. U.S. 2,157,348 (1939); U.S. 2,157,347, both to I.G. Farbenindustrie AG; W. Reppe, and coworkers. *Justus Liebigs Ann. Chem.*, **601**, 81 (1956).
[36] J. Wislicenus. *Justus Liebigs. Ann. Chem.*, **192**, 106 (1878).
[37] N. D. Field, and D. H. Lorenz. *"High Polymers" in Vinyl and Diene Monomers* (Leonard, E. C., ed.), Wiley, New York, **24**, 365 (1970).
[38] S. R. Sandler, and W. Karo. *Polymer Syntheses*, Academic Press, New York, **2**, 214 (1977); D. H. Lorenz. *Encyclopedia of Polymer Science and Technology* (Mark, H. F., Gaylord, N. G., and Bikales, N. M., eds.), Wiley, New York, **14**, 504 (1971).
[39] N. M. Bikales. *Encyclopedia of Polymer Science and Technology* (Mark, H. F., Gaylord, N. G., and Bikales, N. M., eds.), Wiley, New York, **14**, 521 (1971); E. V. Hort, and R. C. Gasman. *Kirk-Othmer Encyclopedia of Chemical Technology* (Grayson, M., ed.), Wiley, New York, 3rd ed., **23**, 937 (1983).
[40] M. Sawamoto, and T. Higashimura. *Makromol. Chem., Macromol. Symp.* (1990), 32 (Invited Lect. Int. Symp. Cationic Polym. Relat. Ionic Processes, 9th, 1989), 131 (1990).
[41] E. J. Goethals, W. Reyntjens, and S. Lievens. *Macromol. Symp.*, 132(*132 International Symposium on Ionic Polymerization*, 1997), 57 (1998).
[42] C. Decker. *Macromol. Symp.*, World Polymer Congress, 37th International Symposium on Macromolecules, **143**, 45 (1999).

3

Polymers

Polymer chemistry starts with a growth of chain by chemical reactions. Polymer reactions proceed with a monomer and with an initiator and/or catalyst. Monomer unit is often used to mean either the chemical repeat unit or the small molecule which polymerizes to give the polymer. Polymer is a particular class of macromolecules with a set of regularly repeated chemical units of the same type, or possibly of a very different type joined end to end, or in more completed ways to form chain molecules. Therefore, polymer materials consist of long chain molecules formed by the chemical composition of monomers. These macromolecules built from smaller molecular subunits namely monomers. The elementary building blocks – monomers – linked together by covalent bonds. Polymers by their common structural feature are the presence of covalently bonded long chains of atoms.

Polymers are accepted due to their properties, excellent chemical resistance, good processability, potential for part consolidation, and assembly simplification due to part consolidation. They are inexpensive possibility of modifying mechanical properties in a wide range by adding fillers and elastomers. They enhance the comfort and quality of life by maintaining hygiene in the modern industrial society from packaging, automobiles, building, etc. Due to their low specific gravity, they occupy high volume fraction despite their relatively low weight fraction [1].

3.1 Classification of Polymers

Polymers are classified in different ways based on:

- Origin – Natural or synthetic. Cellulose, lignin, starch, silk, wool, chitin, natural rubber, proteins, nucleic acids, etc., are naturally occurring polymers. Natural polymers can be modified by chemical reactions as modified natural polymers. Synthetic polymers such as polyethylene, polystyrene, polyurethanes, etc., are synthetic polymers.
- Biodegradable and non-biodegradable.

- Chemical composition – Homo chain and hetero chain based on the type of atom in the main chain. Homo polymers and copolymers – A polymer composed of one type of chemical unit present in the polymer is homopolymer. There is more than one type, it is copolymer that is, a second or third type of monomer is involved in the polymer, the resulting materials are called binary, tertiary etc., copolymers. Different types of arrangement of monomers in the resulting copolymer chains distinguishing amongst others – alternating, statistic, block and graft copolymers.
- Organic and inorganic polymers – carbon, hydrogen, oxygen, nitrogen, halogens, phosphorus and sulphur.
- Polymerization techniques – chain growth polymers are polyethylene, polypropylene, polyvinylchloride, polystyrene, poly(methyl methacrylate), polyacrylonitirle, etc. from chain growth polymerization and polymers such as polyoxymethylene, polyamide, polycarbonate, polyethylene terephthalate, polyphenylene oxide, etc. are from step growth polymerization.
- Thermoplastics and thermosets – Applied heat and pressure such as thermoplastic and thermoset. Polymer materials that can be formed into shapes. Materials that can be shaped or reprocessed more than once are commonly known as thermoplastic polymers. Materials that can be shaped or processed only once are commonly known as thermosetting polymers. Moderate deforming characteristic regain to its original shapes commonly known as elastomers which are properties useful for fibres.
- Polymer architecture – Structure such as linear, branched, network, hyperbranched, dendrimer.
- Structural and functional polymers. – Structural polymers are characterized by and are used because of their good mechanical, thermal, and chemical properties. Hence, they are primarily used as construction materials in addition to or in place of metals, ceramics, or wood in applications. Functional polymers, have completely different property profiles useful in electrical, and in devices for microelectronic, biomedical applications, analytics, synthesis, cosmetics, or hygiene.

Polymer classification is often used on the physico-chemical properties and applications that can be divided into thermoplastics, thermosets, and elastomers. Still the classification is vague, since some polymers can be classified into two groups such as polypropylene is thermoplastic and also synthetic fibres, a copolymer of ethylene and propylene is being thermoplastic and elastomer.

3.2 Evolution of Polymers

Polymers are considered to be an invaluable gift of modern sciences and technology to the mankind. They have become indispensable to our life with their wide range of applications in diverse fields such as packaging, agriculture, food, consumer products, medical appliances, building materials, industry, aerospace materials etc. However, the resistance of polymers to chemical, physical, and biological degradation has become a serious concern when used in areas such as surgery, pharmacology, agriculture and the environment, and as a consequence timeresistant polymeric wastes are becoming less and less acceptable [2].

Naturally, the necessity for polymeric materials satisfying the conditions of biodegradability, biocompatibility and release of low-toxicity degradation products, as an alternative to these existing polymers is apparent. The complexity of polymers is from chemical evolution to nuclei to atoms to molecules to monomers and to polymers. Monomers are comparatively simple and not too difficult to synthesis. Polymers are made from monomers are much complex and their origin is the basic problem. Polymers are useful in daily life as well as in industry. Polymers exhibit superior mechanical properties. Some of the main types of chemical structural repeat units present in the more widely used polymers as follows:

3.3 Recyclable and Non-Biodegradable

3.3.1 Polyolefin

Polyolefin consist of a chain made up of carbon and hydrogen atoms. Polyolefin is an economical polymer and is widely used due to its low cost and versatile properties. They are being hydrophobic in nature. The polyolefins are non-polar, non-porous, low-surface energy that is not receptive to inks, and lacquers without special oxidative pretreatment. They are poor reactivity, dyeability, and hygroscopicity limit. Due to their poor wettability it gives rise to application problems in practice. Surface modification techniques by introducing polar functional groups on the substrate have been exploited to overcome these problems [3, 4]. Therefore, surface free energies lead to an increase in the wettability. Therefore, these polymers require grafting unsaturated polar groups on the backbone [5].

Polyolefin is extensively used as polymer materials due to their low specific gravity, chemical resistance and availability at low cost. Only class of macromolecules can be produced catalytically with precise control of stereochemistry and, to a large extent, of (co)monomer sequence distribution. Polyolefins constitute the most widely used group of commodity polymer. They are prepared by polymerization of simple olefins such as ethylene, propylene, butenes, isoprenes, and pentenes as well as their copolymer [6, 7].

3.3.1.1 Polyethylene (PE)

Polyethylene is the widely used polyolefin polymer. Polyethylene is a generic name for polymers from ethylene monomers. Polyethylene is the simple straight chain hydrocarbon. It is a partial crystalline polymer [8].

Polyethylene has repeated unit of $-CH_2-$ with bond angles of 109°28' and length of 0.154nm. The energy about 80kcal/mole results in polymer. The polymer results in a simple zigzag planar configuration (Figure 3.1) [9–11].

At elevated temperatures oxygen and oxidizing agents attack polyethylene. The chemical degradation results in physical changes and ultimately in the destruction of the useful properties [12, 13]. The double bonds in polyethylene are normally present to the extent of, only one per molecule. It has little effect on physical properties [14–17].

Polyethylene is described as low, high and linear low by its specific gravity, strength, molecular weight, crystallinity, and melt flow or rheological characteristics. The value of long chain branching in linear polymers with superior physical properties is available with high density PE (HDPE) [18] or linear low density PE (LLDPE) [19]. LLDPE with superior properties would replace LDPE in many applications. Some of the properties of polyethylene are given in Table 3.1.

Polyethylene is nonpolar, and finds application in emerging technologies. It is functionalized with various monomers which includes maleic anhydride, diethyl maleate with free radical initiators. Terminal unsaturation in PE can

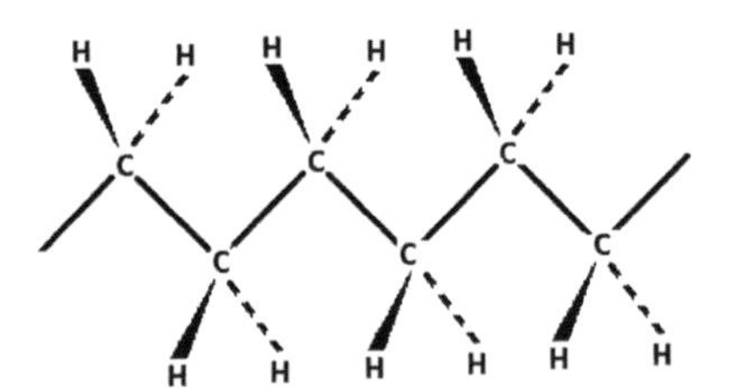

Figure 3.1 Covalently bonded hydrocarbons.

Table 3.1 Typical properties of polyethylene

Polyethylene – Repeat Unit	$\left[CH_2 - CH_2 \right]_n$	
Property	Unit	Value
Specific gravity		0.910–0.970
Glass transition temperature	°C	–120
Melt temperature	°C	115–137
Thermal conductivity	W/mK	0.33–0.52
Heat deflection temperature	°C at 445kPa	42–85
Heat of combustion	KJ/g	46.5
Thermal linear expansivity	$10^{-5}K^{-1}$	11–20
Specific heat capacity	kJ $kg^{-1}K^{-1}$	1.90–2.31

increase the amount of crosslinking that occurs during peroxide modification. However, there is discrepancy with the exact mechanism by which the terminal unsaturation is involved [20, 21].

Polyethylene has several practical properties such as low cost, excellent electrical insulation, good processability and resistance to chemicals [22, 23]. It has also high mechanical strength, low brittle temperature, flexibility and outstanding electrical properties [24]. Polyethylene has good mechanical properties to use in the industry [25].

Polyethylene has drawbacks which include low environmental stress cracking resistance (ESCR) and poor compatibility with various additives which have restricted its use in certain purposes. Therefore, properties such as environmental stress crack resistance and optics tend to be application specific measures. With increase in molecular weight, it has better tensile and environmental stress cracking resistance [26, 27].

3.3.1.2 Polypropylene (PP)

Polypropylene have the repeated unit of $–CH_2–CH(CH_3)–$ [28–31]. Polypropylene encompasses its versatility in physical properties, regarding its ease of manufacturing and its use in many applications [32, 33]. Some of the properties of polypropylene is given in Table 3.2.

Polypropylene (PP) exhibits high strength, high wear and chemical resistance, low water vapor and gas permeability, and good dielectric characteristics. However, PP is brittle, and its frost resistance is poor (–15°C) [34]. The poor transparency and brittleness restricts its application in many field particularly transparency requirement products such as medical and personal care products. The material was developed by using Ziegler type catalysts [35].

Table 3.2 Typical properties of polypropylene

Polypropylene – Repeat Unit	$\left[-CH(CH_3)-CH_2- \right]_n$	
Property	**Unit**	**Value**
Specific gravity		0.900
Glass transition temperature	°C	–16
Melt temperature	°C	168–172
Thermal conductivity	W/m K	0.24
Heat deflection temperature	°C at 445kPa	115
Heat of combustion	KJ/g	46
Thermal linear expansivity	$10^{-5}K^{-1}$	6–10
Specific heat capacity	$kJ\ kg^{-1}\ K^{-1}$	1.93

3.3.2 Polystyrene

Polystyrene (PS) is most widely used commercial polymer. Polystyrene is a synthetic aromatic polymer made from the monomer styrene. Polystyrene can be solid or foamed. PS is a relatively low cost polymer, having some excellent properties. PS is a hard, rigid, clear, and amorphous polymer. PS is in a solid (glassy) state at room temperature and flows if heated above about 100°C, this change in temperature is known as glass transition temperature. It becomes rigid again when cooled. It can be naturally transparent. Some of the physical properties are given in Table 3.3.

Table 3.3 Typical properties of polystyrene

Polystyrene – Repeat Unit	$\left[-CH(C_6H_5)-CH_2- \right]_n$	
Property	**Unit**	**Value**
Specific gravity		1.06
Glass transition temperature	°C	95–100
Melt temperature	°C	210–220
Thermal conductivity	W/m K	0.16
Heat deflection temperature	°C at 445kPa	82
Heat of combustion	KJ/g	42.2
Thermal linear expansivity	$10^{-5}K^{-1}$	6–8
Specific heat capacity	$kJ\ kg^{-1}\ K^{-1}$	1.20

Polystyrene is used in a wide range of commercial applications as packaging material, injection moulded parts, and UV screening agents [36, 37]. It has good thermal and radiation resistant properties and available in a wide range of formulations. It is somehow poor barrier to oxygen and water vapour. It has a relatively low melting point. However, it has a prominent limitation in its brittleness. It is used in protective packaging, containers, disposable cutlery. Application includes in items such as toys, cups, office supplies, packaging materials, and surface coatings. Use of PS is limited because of its susceptibility to degradation from UV radiation, chemical attack from aromatic, and chlorinated hydrocarbon may cause problem in industrial applications.

3.3.3 Polyvinylchloride (PVC)

Polyvinylchloride (PVC) is halogen-substituted hydrocarbon. The structural defects in PVC such as tertiary chlorine at branched carbons, allylic chlorines adjust to internal unsaturation, oxygen containing groups, end groups, etc. are of practical importance despite their low concentration. They affect the colors, thermal stability, crystallinity, processing, and mechanical properties of finished materials [38, 39].

Polyvinylchloride have the repeated unit of $-CH_2-CH(Cl)-$ [40, 41]. It has high level of combustion resistance. Good flame retardancy of PVC is due to its high chlorine content. However, it releases high levels of smoke and toxic gases during burning [42–45]. Smoke suppressants are added to PVC during its processing being PVC has been increasingly used to produce building materials. It has excellent mechanical properties and economical. Some of the physical properties are given in Table 3.4.

Table 3.4 Physical properties of PVC

Polyvinylchloride – Repeat Unit	$\left[-CH(Cl)-CH_2- \right]_n$	
Property	Unit	Value
Specific gravity		1.380–1.410
Melt temperature	°C	245–265
Limiting oxygen index (LOI)		0.206
Glass transition temperature	°C	87
Melt temperature	°C	175–212
Thermal conductivity	W/m K	0.15–0.16
Heat of combustion	KJ/g	19.9
Thermal linear expansivity	$10^{-5}K^{-1}$	5–25
Specific heat capacity	$kJ\ kg^{-1}\ K^{-1}$	1.05

Polyvinylchloride uses in medical and personal care products with wide processing window. It is economic material with growing recycling concern [46]. This material has found applications in construction, the toy industry, and even blood and plasma conditioning [47] and also in packaging materials exposed to ionizing radiation. Functional additives permit the generation of PVC into rigid and flexible products, useful in designed engineering application [48]. It has outstanding chemical resistance to wide range of corrosive fluids. Flexibility from plasticization occurs with the polar groups of PVC and replaces polymer-polymer interactions with polymer-plasticizer interactions thus shielding polymer chains from interacting with other [49–51].

3.3.4 Poly(methyl methacryalate) (PMMA)

Poly(methyl methacrylate) is a vinylidene polymer. It is white powder and stable polymer. This polymer is combustible, and incompatible with oxidizing agents. PMMA is optically clear through the visible wavelength range. It has good mechanical properties and high weather properties. However, PMMA is high sensitive to electron radiation [52, 53].

Poly(methyl methacryalate) is widely used as materials for biomedical implants, barriers, membranes, micro lithography, and optical applications. Poly(methyl methacrylate) (PMMA) is an important thermoplastics. PMMA is a flexible coil segment. It will decompose at about 220°C and liberates large amounts of monomer. It is used widely as a building material and consumer product. It has excellent transparency and good weathering resistance. It can burn to release heat, smoke and toxic gases. PMMA is a transparent polymer with high glass transition temperature [54]. Some of the properties of poly(methyl methacrylate) are given in Table 3.5.

3.3.5 Polyamide

Polyamide is an engineering polymer. It has high strength, wear and heat resistance properties. It is however, relatively expensive. It has poor impact strength and moisture resistance properties. Polyamides are important engineering thermoplastics due to their excellent strength and stiffness, low friction, and chemical resistance. However, polyamides are highly notch sensitive. They are often ductile in the unnotched state, but fail in a brittle manner when notched. The difference between unnotched and notched Izod impact strength indicates that nylons are fairly resistant to crack initiation but have only modest resistance to crack propagation. Some of the properties of polyamide are given in Table 3.6.

Table 3.5 Properties of poly(methyl methacrylate)

Poly(methyl methacrylate) – Repeat Unit

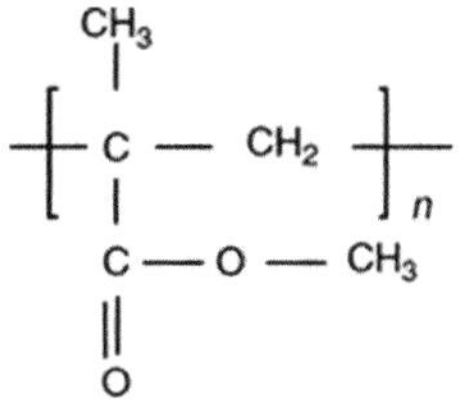

Property	Unit	Value
Specific gravity		1.17–1.22
Glass transition temperature	°C	90–105
Melt temperature	°C	
Thermal conductivity	W/m K	0.19
Heat deflection temperature	°C at 445kPa	93
Heat of combustion	KJ/g	26.2
Thermal linear expansivity	$10^{-5}K^{-1}$	4.5
Specific heat capacity	kJ kg^{-1} K^{-1}	1.39

Table 3.6 Physical properties of polyamide

Polyamide – Repeat Unit

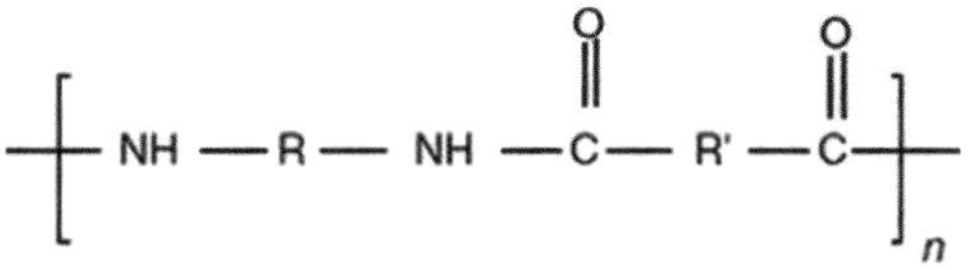

Property	Unit	Value
Specific gravity		1.14
Glass transition temperature	°C	180–265
Melt temperature	°C	
Thermal conductivity	W/m K	0.25–0.31
Heat deflection temperature	°C at 445kPa	245
Heat of combustion	KJ/g	31.9
Thermal linear expansivity	$10^{-5}K^{-1}$	6–9
Specific heat capacity	kJ kg^{-1} K^{-1}	1.6–1.7
Glass transition temperature T_g	°C	–16
Melting temperature T_m	°C	168–172
Limiting oxygen index (LOI)		0.243–0.287

Owing to the possibility of hydrolytic degradation polyelectrolyte affects the presence of aggregates and apparatus problems like corrosion or decomposition of membranes molecular weight determinations may be difficult or even impossible. Semi-crystalline polyamides are biodegraded only extremely slowly [55, 56].

3.3.6 Polycarbonate

Polycarbonate (PC) is an engineering polymer with several important characteristic properties such as high toughness, high continuous working temperature, high modulus, and transparency. However, major drawbacks of PC include high melt viscosity and notch sensitivity. The disadvantages of PC can be overcome through blending with various polymers, of these; acrylonitrile–butadiene–styrene (ABS) copolymer is the most popular [57, 58]. Some of the properties of polycarbonate are given in Table 3.7.

Polycarbonate is divided into a linear and a branched polymer according to its structure and these polymers show a different rheological behaviour [59]. A variety of new applications have been created by tailoring the base polycarbonate polymer with modification that enhances the end-use properties of the polymer [60].

Polycarbonate divides into linear and branched polymer according to its structure [61]. Branched polymers are used for the applications characterized by extensional flow fields in the molten state such as extrusion coating, blow moulding, foam extrusion, and melt phase thermoforming [62, 63]. Generally, branched polymer exhibits higher molecular weight than linear polymer, therefore it exhibits lower flowability.

3.3.7 Poly(ethylene terephthalate) (PET)

Poly(ethylene terephthalate) (PET) is an aromatic polyester. It used in the packaging industry particularly as food or beverage containers. Poly(ethy lene terephthalate) (PET) is widely used in industry as fibres for

Table 3.7 Properties of polycarbonate

Polycarbonate – Repeat Unit	$[-O-C_6H_4-C(CH_3)_2-C_6H_4-O-C(=O)-]_n$	
Property	Unit	Value
Specific gravity		1.2
Glass transition temperature	°C	145
Melt temperature	°C	230
Thermal conductivity	W/m K	
Heat deflection temperature	°C at 445kPa	138
Heat of combustion	KJ/g	30.8
Thermal linear expansivity	$10^{-5}K^{-1}$	
Specific heat capacity	$kJ\ kg^{-1}\ K^{-1}$	

Table 3.8 Physical properties of polyethylene terephthalate

Polyethylene terephthalate – Repeat Unit	$-[(CH_2)_2-O-C(=O)-C_6H_4-C(=O)]_n-$	
Property	**Unit**	**Value**
Specific gravity		1.290–1.400
Glass transition temperature	°C	75–80
Melt temperature	°C	255–265
Thermal conductivity	W/m K	0.14
Heat deflection temperature	°C at 445kPa	38
Heat of combustion	KJ/g	21.6
Thermal linear expansivity	$10^{-5}K^{-1}$	10
Specific heat capacity	kJ kg^{-1} K^{-1}	1.01
Limiting oxygen index (LOI)		0.206

cloths, base films of magnetic tapes, floppy discs, photo recording medium or photographic film, etc. [64, 65].

Poly(ethylene terephthalate) (PET) is one of the most interesting polymer due to its molecular structure can be varied very widely. The high melting points of the polyesters obtained by condensation of glycols with terephthalic acid and other symmetrical aromatic dibasic acids [66, 67] are of interest due its superior chemical, physical, mechanical, and (oxygen and carbon dioxide) barrier properties. Some of the properties of polyethylene terephthalate are given in Table 3.8.

When the molten polymer quench to room temperature, it solidifies in a state of disorder called amorphous state. This amorphous material subsequently heated to temperature above about 100°C, crystallization occurs and not oriented, partial crystalline material is formed. The degree of crystallinity varies depending on the temperature and heat treatment time. Stretching the unoriented polymer produces material of varying crystallinity and molecular orientation.

3.3.8 Poly(butylene terephthalate) (PBT)

Poly(butylene terephthalate) (PBT) is an engineering polymer due to its good mechanical strength and toughness, strong dimensional stability, thermal resistance, and processing advantages. However, PBT has poor notched impact strength, so it is necessary to significantly improve its notched impact strength to meet more field requirements. The rate of crystallization of PBT is high that crystallization can be avoided by controlled quenching of the melt [68]. Some of the properties of poly(butylene terephthalate) are given in Table 3.9.

Table 3.9 Physical properties of poly(butylene terephthalate) (PBT)

PBT – Repeat Unit	$\left[-\overset{O}{\overset{\|}{C}}-C_6H_4-C-O-(CH_2)_4-O- \right]_n$	
Property	Unit	Value
Specific gravity		
Glass transition temperature	°C	
Melt temperature	°C	255–265
Thermal conductivity	W/m K	
Heat deflection temperature	°C at 445kPa	
Heat of combustion	KJ/g	
Thermal linear expansivity	$10^{-5}K^{-1}$	
Specific heat capacity	$kJ\ kg^{-1}\ K^{-1}$	

3.3.9 Poly(ether ether ketone) (PEEK)

Poly(ether ether ketone) (PEEK) is an engineering polymer with properties such as high temperature resistance and chemical resistance. The polymer has high mechanical properties and self-lubricating. Therefore, it is widely used in chemical, mechanical, aeronautic, electronic and nuclear industries. With high performance, the polymer also has high glass transition temperature with high melting temperature [69]. It has good thermal stability, chemical resistance, and excellent mechanical properties [70–72]. PEEK uses widely for high performance composites. Some of the properties of polyether ether ketone are given in Table 3.10.

Table 3.10 Properties of polyetherether ketone

Polyether ether ketone – Repeat Unit	$\left[-O-C_6H_4-O-C_6H_4-\overset{O}{\overset{\|}{C}}-C_6H_4- \right]_n$	
Properties	Unit	Value
Specific gravity		1.31
Glass transition temperature	°C	
Melt temperature	°C	
Thermal conductivity	W/mK	
Heat deflection temperature	°C at 445kPa	160
Heat of combustion	KJ/g	
Thermal linear expansivity	$10^{-5}K^{-1}$	
Specific heat capacity	$kJkg^{-1}K^{-1}$	

It exhibits very good thermal stability [73], high chemical resistance and excellent mechanical properties [74]. The thermal stability have indicated is stable in nitrogen up to 400°C, holding for 15 min [75].

3.3.10 Poly(oxy methylene)

Poly(oxy methylene) is also known as polyacetal. It is a crystalline polymer with high crystallinity, hardness, strength, and stiffness. It has a good lubricity and heat resistance. It also has low coefficient of friction with good chemical resistance to most solvents [76, 77]. POM is useful in mechanical, automotive, plumbing, appliance, industrial, and electrical components. It replaces many metals in various end-use applications [78].

The end groups of poly(methylene oxide) are to be modified in order to prevent unzipping. To increase the stability, the inclusion of copolymerization with another monomer such as ethylene oxide in the main chain use to modify the polymer. Solvent with a similar solubility of its monomer will cause swelling. There is no solvent at room temperature. At high temperature, there are solvents such as phenol, aniline, and benzyl alcohol, may dissolve the polymer. Some of the properties of poly(oxy methylene) are given in Table 3.11.

Poly(oxymethylene) have repeated unit of $-CH_2-O-$ [79–81]. Generally POM is good resistance to thermal and oxidative degradation. Properties will have little impact from humidity. Low gas and vapour permeability. Polyacetal shows little effect on properties from humidity. Polyacetal have excellent fatigue resistance and resistance to creep. They also have a low coefficient of

Table 3.11 Properties of poly(methylene oxide)

Poly(oxy methylene) – Repeat Unit		$\left[-\overset{H}{\underset{R}{C}}-O- \right]_n$
Properties	Unit	Value
Specific gravity		1.42
Glass transition temperature	°C	–76
Melt temperature	°C	165–175
Thermal conductivity	W/m K	0.23
Heat deflection temperature	°C at 445kPa	165
Heat of combustion	KJ/g	16.9
Thermal linear expansivity	$10^{-5}K^{-1}$	10
Specific heat capacity	$kJ\ kg^{-1}\ K^{-1}$	1.47

friction and excellent abrasion resistance. POM is useful in electrical applications particularly in non-critical applications. Not suitable for application at high frequencies (1 GHz) or applications involving high electric stress above 70°C.

3.4 Non-Biodegradable and Non-Recycleable Polymers

3.4.1 Phenol Formaldehyde

Phenol formaldehyde commercially called as Bakelite. They are polymers of curing nature. Phenols, substituted phenols, formaldehyde are typical comonomers. It undergoes step growth polymerization and simple structure illustrated in Figure 3.2. They are useful in construction materials, electronics, aerospace, moulded parts, etc. The properties such as toughness, temperature resistance, low void content, chemical resistance, corrosion inhibition lead to many home and industrial applications [82–84]. Some of the physical properties are given in Table 3.12.

3.4.2 Epoxy

Epoxy resins are one of the most important thermosetting polymers and have wide use as structural adhesives and matrix resin for fibre composites, but their cured resins have one drawback. They are brittle and have poor resistance to crack propagation. Epoxy materials are cross-linked polymers after curing. Uncured base resins are diglycidyl ether of bisphenol A and are not cross-linked. To cross-link, this polymer requires diluents, modifiers, fillers, colorants and dyes and other additives [85, 86]. The properties depend on the

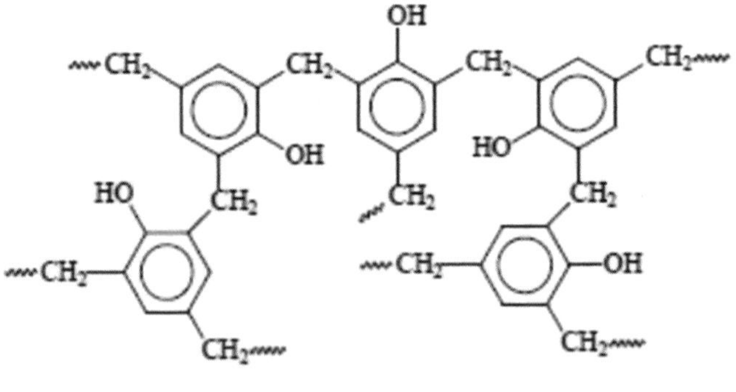

Phenol formaldehyde

Figure 3.2 Structure of phenol formaldehyde.

Table 3.12 Properties of phenol formaldehyde

Properties	Unit	Value
Specific gravity		1.370–1.460
Dissipation factor		0.03–0.07
Dielectric constant		4.0–7.0
Heat deflection temperature	°C	125–172
Hardness	Rockwell	M 96–M 120
Type of polymer		Thermoset

formulation and processing. They have good weather resistance and durability. It has better mechanical properties than any other castable polymer. It is skin sensitizers. It discolors when exposed to ultra-violet rays. Initially formation of low molecular with epoxy end groups in two-stage process. Ethylene diamine reacts with epoxy end groups on low molecular weight polymer.

The use of a wide range of temperature of curing agents with good control over the degree of cross-linking and availability of the resin ranging from low viscous liquid to tack free solid etc The simple structure of epoxy polymer is mentioned in Figure 3.3. Epoxides react with amines, carboxylic acids, anhydrides, etc. to form polymers display a range of characteristics from tough, durable to soft products, and adhesive.

The reactive species of two or more epoxy groups undergoes reaction and form epoxy polymers with curing agents. A reactive epoxide or a mixture of reactive epoxides, and also its admixture with curing agents or agents, is commonly referred as epoxy resin. The reactive epoxides are cured by reactions with a di or higher functional amine, carboxylic acid, anhydride, or cyanoguanidine. They are useful in protecting appliances, automotive

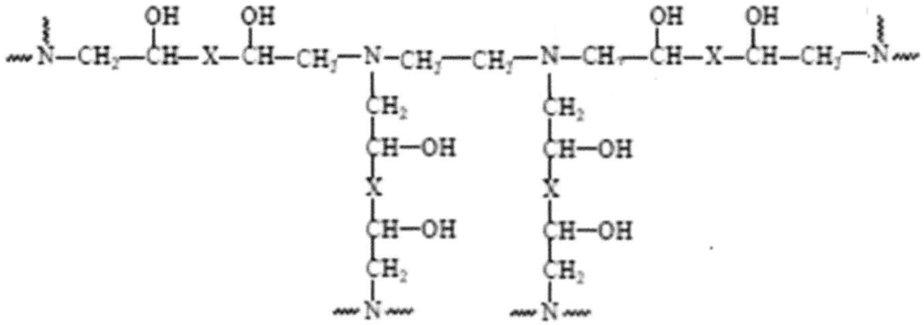

Figure 3.3 Structure of epoxy polymer.

primers; pipes etc. and are also useful in encapsulation of electrical and electronic devices. It is used as adhesive and bonding agents for dental use.

3.4.3 Urea and Melamine Formaldehyde

Urea or melamine as the amino component and formaldehyde is used commonly to the resulting polymer named as urea-formaldehyde and melamine-formaldehyde. A mixture of reactive oligomers prepared by the condensation of an amine monomer with an aldehyde usually is named as amino resin. Oligomers present in urea – formaldehyde and melamine (Figure 3.4) – formaldehyde resins (Figure 3.5) are:

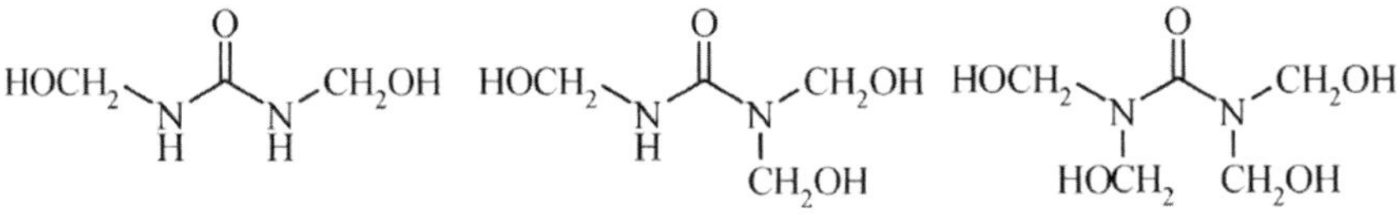

N,N'-bis(hydroxymethyl)urea tris(hydroxymethyl)urea tetrakis(hydroxymethyl)urea

Figure 3.4 Oligomers present in urea formaldehyde.

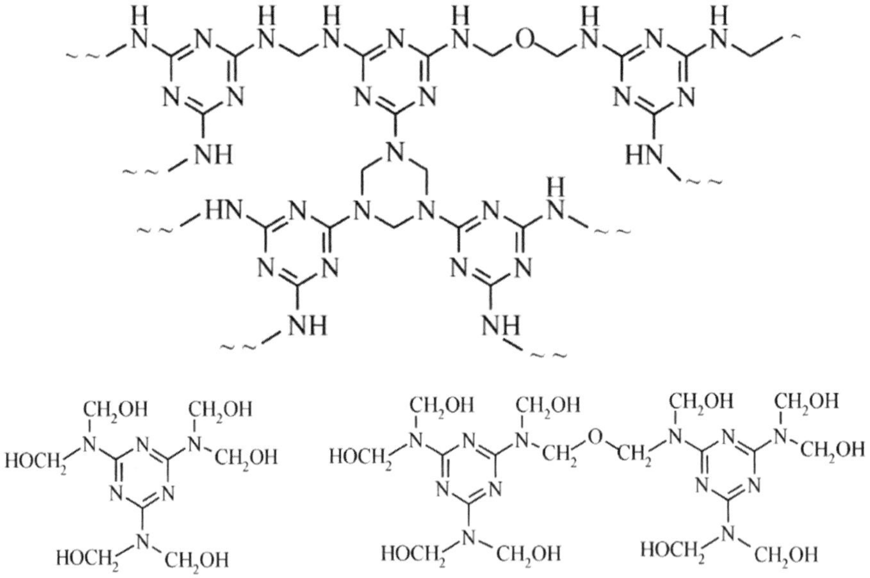

Figure 3.5 Oligomers present in melamine formaldehyde.

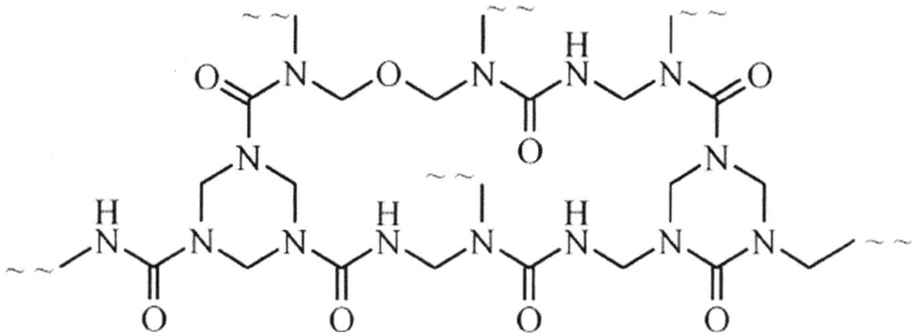

Figure 3.6 Network structure of urea-formaldehyde.

The amino monomer with formaldehyde, the cured resin should be named as amino polymer. It is usually a network polymer (Figure 3.6). Polycondensation takes place at the hydroxymethyl groups with formation of network structure. The polymer of melamine – formaldehyde can be as a subclass of polytriazines.

3.4.4 Unsaturated Polyesters

Unsaturated polyesters are economical, low density, good corrosion resistance, and high strength to weight ratio. This polymer is also known as thermoset polyesters. It is used as a resin component for composites in the building industry such as dome light crowns, in transportation sector in boats and ships building, tanks, tubes, vessels, and others, and in electrical industry for cable distribution cupboards. They are highly flammable and produce large quantities of smoke and toxic gases when burns [87, 89]. Polymerization involves a radical polymerization between a pre-polymer that contains unsaturated groups and styrene [90, 91].

3.5 Copolymers

Copolymers are macromolecules. It contains more than one type of monomer unit within the polymer chain. The properties vary in wide range by copolymerization. Many commercial polymers contain small amounts of other monomers. The operating condition controls the architecture of the copolymer. The control is also from the chemical composition and physical state of catalyst used. The copolymerization carried out in the liquid monomer, in a solvent or in aqueous emulsion. Solvents with low chain transfer results in high molecular mass.

3.5.1 Acrylonitrile-Butadiene-Styrene (ABS)

ABS (acrylonitrile/polybutadiene/styrene) graft copolymers are widely used in many industrial applications. Acrylonitrile-butadiene-styrene (ABS) is an engineering polymer. It consists of an amorphous, hetero phasic polymer with very good mechanical properties, especially high impact resistance. Typically, ABS consists of a styrene/acrylonitrile continuous phase partially grafted to a dispersed butadiene phase. The latter acts as an impact modifier, and imparts excellent mechanical properties to the material. ABS forms from three different monomers. Acrylonitrile contributes heat resistance, high strength and chemical resistance, butadiene contributes impact strength, toughness and low temperature properties retention and styrene contributes gloss, processing ability and rigidity. Properties of ABS modify by varying the relative proportion of the basic components, the degree of grafting and molecular weight. It is normally an opaque polymer with each phase having different refractive index. Being a versatile polymer, it tailors to meet specific product requirements. Some of the physical properties are given in Table 3.13.

Rubber particles are much small and contain styrene-acrylonitrile copolymer inclusion. The average particle size and particle size distribution within the rubber phase have significant effects on the overall balance of properties. The properties include strength, toughness, appearance, etc. A large number of rubber particles typically enhance toughness while lowering appearance. The rigid phase chain length can also have significant effect on ABS.

The longer the chain, the higher the strength includes impact/ductility, while lowering the flow during processing. The ratio of styrene and acrylonitrile to the butadiene affects the flow and impact balance of the ABS.

Table 3.13 Acrylonitrile-Butadiene-Styrene

Acrylonitrile-Butadiene-Styrene – Repeat Unit		
Properties	Unit	Value
Specific gravity		1.030–1.060
Melt temperature	°C	230–250
Glass transition temperature	°C	88–120
Water absorption	%	0.4
Vicat softening temperature	°C	92–101

Increase in butadiene content will increase the impact and toughness with scarification of the flow. The balance between the flow and impact properties of ABS material is the basic characteristic. The main applications of ABS are in automotive industry, telecommunications, business machines, and electric/electronic casing. The wide range of ABS application is due to both the properties and price which are intermediate between the lower priced commodity thermoplastics and the more expensive high performance-engineering polymer.

3.6 Biocompatible and Biodegradable Polymers

Man-made polymers that do biodegrade tend to have structures similar to those found in naturally occurring polymers. The microbes produce enzymes that may not discriminate between polymers of similar structures. The biodegradable polymers include poly(vinyl alcohol) which is the only carbon chain polymer. A few synthetic polymers, such as polylactide [PLA], polycaprolactone and polyglycolide [PGA] are known for their usefulness in medicine [92–94].

These biocompatible and biodegradable polymers can form the basis of an environmentally preferable, sustainable alternative to current materials based exclusively on petroleum feed stocks [95, 96]. The availability of biocompatible and biodegradable polymers has promoted major advances in the biomedical field [97]. However, the cost of this process is high due to the complexity of the process. These materials have advantages in terms of their sustainability and lifecycle, especially as relating to carbon-based polymeric materials. Biopolymers are generally capable of being disposed in safe and ecologically sound ways through disposal processes (waste management) such as composting, soil application, and biological wastewater treatment. Single use, short-life, disposable products can be engineered to be bio-based and biodegradable.

3.6.1 Poly(vinyl alcohol) (PVA)

Polyvinyl alcohol (PVA) is a unique synthetic polymer and credits wide spectrum of biomedical applications such as soft contact lenses [98], implants [99], and artificial organs [100]. Poly(vinyl alcohol) (PVA) is non-toxic, non-carcinogen, good biocompatibility and desirable physical properties such as rubbery or elastic nature. It has a characteristic of high degree of swelling in aqueous solution. Some of the physical properties are given in Table 3.14.

Table 3.14 Physical properties of poly(vinyl alcohol)

Poly(vinyl alcohol) – Repeat Unit		$\left[CH_2\underset{OH}{\underset{|}{C}}H \right]_n$
Properties	Unit	Value
Specific gravity		1.27–1.31
Melt temperature	°C	212–235
Glass transition temperature	°C	75–85
Specific heat capacity	KJ/Kg C	1.67
Thermal conductivity	W/mK	0.02–0.4

The chemical and physical properties of PVA have led to its broad industrial use. Chemical cross linking of PVA leads to feasible routes for the improvement of the thermal stability and physical properties.

3.6.2 Poly(ethylene glycol) (PEG)

Polyethylene glycol is a linear polymer. It is a free flowing powder [101, 102]. It is soluble in polar solvents like dimethylformamide, methanol, and water but insoluble in diethyl ether or isopropanol. It is used as a reference which is a biocompatible polymer for biomedical field due to its unique properties of chemical stability, solubility in organic and aqueous media, non-toxicity, low immunogenicity and antigenicity [103].

PEGs have become very popular in organic synthesis since they are non-volatile, recyclable, stable to acid, base, and high temperature, and available in high quantities at low prices. Furthermore, their well-known low toxicity makes them greener versions of conventional halogenated solvents [104]. Some of the physical properties are given in Table 3.15.

Poly(ethylene glycol) (PEG) is the most commonly applied non-ionic hydrophilic polymer with stealth behaviour. Furthermore, PEG reduces the tendency of particles to aggregate by steric stabilization, thereby producing formulations with increased stability during storage and application. Furthermore, their well-known low toxicity makes them greener versions of conventional halogenated solvents. Poly(ethylene oxide) (PEO), also referred to as poly(ethylene glycol) (PEG), is the reference biocompatible polymer used in the biomedical field as a result of its unique properties, namely, its chemical stability, solubility in both organic and aqueous media, nontoxicity, low immunogenicity, and antigenicity [105].

Table 3.15 Physical properties of poly(ethylene glycol)

Polyethylene glycol – Repeat Unit	$-[CH_2 - CH_2 - O]_n-$	
Properties	Unit	Value
Specific gravity		1.127–1.210
Melt temperature	°C	37–62
Glass transition temperature	°C	–15
Surface tension	mN/m	33.8
Refractiveindex		1.46

3.6.3 Polylactide or poly(lactic acid) (PLA)

Polylactide (PLA) or poly(lactic acid) is valuable in biomedical and pharmaceutical applications due to its biodegradable nature. It is a typical biodegradable polymer made from bioresources, which has extensively been used in many industrial fields such as automotive, electrical, and medical industries. However, the brittleness of PLA impedes its development for large-scale commercial use [106]. PLA has various attractive properties, such as high rigidity, biodegradability, and biocompatibility. It will be required to overcome the following defects; (1) slow crystallization, (2) poor heat resistance, and (3) mechanical brittleness [107]. Some of the physical properties are given in Table 3.16.

This polymer decomposes rapidly and completely in a typical compost environment which makes it an ideal replacement for non-degradable polymers in numerous applications like yard waste bag and food containers, etc.

Table 3.16 Physical properties of polylactic acid

Polylactic acid–Repeat Unit	$-(CH(CH_3)-C(=O)-O)_n-$	
Properties	Unit	Value
Specific gravity		1.248–1.290
Melt temperature	°C	164.3
Glass transition temperature	°C	62.7
Decomposition temperature	°C	414.7
Activation energy E_a	KJ/mol	87.0
Change in enthalpy(melting)	J/g	0.4

[108, 109]. Moreover, PLA is a biodegradable polymer with a relatively lower cost than other biodegradable polymers. However, commercialized PLA has low crystallinity and a slow rate of crystallization, which limits its use in plastic or film applications.

PLA has applications in bio-absorbable medical devices as bone fixation plates, screws and rods used in orthopaedic and oral surgeries [110–112]. However, polymer exhibits brittle fracture behaviour under impact loading conditions. Therefore toughening of PLA becomes one of the most important issues in the field of biopolymer engineering [113–124].

3.6.4 Poly(caprolactone)

Poly(caprolactone) (PCL) is a semi-crystalline polymer. It has good drug permeability with very slow degradation rate. There are two enantiomers dextro and lavo. They differ in their properties. The dextro polymer is amorphous and has poor drug permeability with high degradation rate [125, 126]. A polymer has to get specifically tailored in order to meet various requirements such as controlled release of different drugs. These polymers are useful as important class of biocompatible and biodegradable material for controlled drug release [127–129]. Some of the physical properties are given in Table 3.17.

3.6.5 Poly(glycolic acid) (PGA)

Poly(glycolic acid) is a simple polyester poly(α-hydroxy acid). It has been the first bioresorbable synthetic polymer. It is used in biomedical applications [130, 131]. This polymer has hydrolytic instability. Synthesis of PGA is polymerized from α hydroxyl acetic acid commonly known as glycolic acid. Glycolic acids form cyclic dimers on mild heating. It is subjected to a

Table 3.17 Physical properties of poly(caprolactone)

Poly(caprolactone) – Repeat Unit		$\left[(CH_2)_6O - \overset{O}{\overset{\|}{C}} \right]_n$
Properties	Unit	Value
Degree of crystallinity	%	69
Melt temperature	°C	58
Glass transition temperature	°C	–72.15
Heat of fusion ΔH_f	KJ/mol	8.9
Water content	%	<1

Table 3.18 Physical properties of poly(glycolic acid)

Poly(glycolic acid) – Repeat Unit		$-[O-CH_2-CO]_n-$
Properties	Unit	Value
Specific gravity		1.69
Degree of crystallinity	%	46–52
Melt temperature	°C	222
Glass transition temperature	°C	45
Decomposition temperature	%	254
Heat of fusion ΔH_f	KJ/mol	8.1

catalytic ring opening the dimers polymerize to form high molecular weight PGA. It is high crystalline polymer. This polyester degrades hydrolytically in the body and generates non-toxic products [132–134]. Some of the physical properties are given in Table 3.18.

3.7 Summary

- Man-made polymers are originally developed for their durability and resistance to all forms of degradation including biodegradation.
- Polymers are accepted worldwide due to their properties and enhance the comfort and quality in modern society.
- Polymers are low density and occupy a high volume seems in the environment as litter and contribution to landfill depletion.
- Non-degradable polymers are strong, light weight, economical, and energy efficient.
- Biodegradable polymers may never replace the major commodity polymers.
- Biodegradable polymers will offer viable waste management options that are not readily or economically recoverable for recycling.

References

[1] M. Thayer. *Chem. Eng. News* **7** (1989).

[2] M. Vert, I. D. Santos, S. Ponsart, N. Alauzet, J. L. Morgat, and J. Coudane. *Polym. Int.* **51**, 840–844 (2002).

[3] O. S. Kolluri. *In Handbook of Adhesive Technology*, (A. Pizzi and K. L. Mittal) Marcel Dekker, New York (1994).

[4] K. Chen. Y. U. Ikada. In *Contact Angle, Wettability and Adhesion.* (Mittal K. L. Ed.). VSP, Utrecht, 865 (1993).

a. I. Mondragon, M. Gazelumendi, and Nazabal. *J. Polyme Eng. Sci*. **28**, 1126 (1998).

[5] I. K. Mahta, S. Kumar, G. S. Chauhan, and B. N. Mishra. *J. Appl. Polym. Sci.* **41**, 1171 (1990).

[6] R. Greco, G. Maglio, and P. Muste. *J. Appl. Polym. Sci.* **33**, 2531 (1987).

[7] C. Vasile, and R. B. Seymour (Eds). H*andbook of Polyolefins*. Marcel Dekkar, New York, NY (1993).

[8] A. Abe, R. L. Jernigan, and P. J. Flory. *J. Am. Chem. Soc*. **88**, 631 (1966).

[9] P. J. Flory. *J. Polym. Sci. Polym. Phys. Ed*. **11**, 621 (1973).

[10] E. V. Thompson. in *Encyclopedia of Polymer Science and Engineering*, Vol. 16, eds H. F. Mark, N. M. Bikales, C. G. Overberger, G. Menges, and J. I. Kroschwitz (Wiley-Interscience, New York, NY), 711–737 (1985).

[11] A. Abe, R. L. Jernigan, and P. J. Flory *J. Am. Chem. Soc*. **88**, 631 (1966).

[12] C. A. Hoeve *J. Chem. Phys.* **35**, 1266 (1961).

[13] Y. Yang, J. Zhang, Y. Zhou, G. Zhao, C. He, Y. Li, et al. *J. Phys. Chem. C* **114**:3701–3716 (2010).

[14] I. Mondragon, M. Gazelumendi, and J. Nazabal. *Polym. Eng. Sci.* **28**, 1126 (1998).

[15] I. K. Mahta, S. Kumar, G. S. Chauhan, and B. N. Mishra. *J. Appl. Polym. Sci.* **41**, 1171 (1990).

[16] R. Greco, G. Maglio, and P. Musto. *J. Appl. Polym. Sci*. **33**, 2531 (1987).

[17] K. Ziegler, and Brennstoff. *Chemistry B-C* 35, 321; German Patent No. 878, 560, (1953).

[18] F. J. Karol. *Macromol. Symp*. **563**, 8–9 (1995).

[19] A. Smedberg, T. Hjertberg and B. Gustafsson. *Polymer* **38**, 4127 (1997).

[20] P. Hendra, A. Peacock, and H. Willis. *Polymer* **28**, 705 (1987).

[21] Y. Yang, J. Zhang, Y. Zhou, G. Zhao, C. He, Y. Li et al. *J. Phys. Chem. C* **114**, 3701–3716 (2010).

[22] C. Zhao, H. Qin, F. Gong, M. Feng, S. Zhang, and M. Yang. *Polym. Degrad. Stab.* **87**, 183–189 (2005).

[23] C. Vasile, and R. B. Seymour (Eds). *Handbook of Polyolefins*. Marcel Dekkar, New York, (1993).
[24] I. M. Ward. *Mechanical Properties of Solid Polymers*, 2nd Edn. John Wiley, New York (1983).
[25] N. S. Vijayalakshmi, and R. A. N. Murthy. *J. Appl. Polym. Sci.* **44**, 1377–1382 (1992).
[26] A. Sharif, N. Mohammadi, and S. R. Ghaffarian. *J. Appl. Polym. Sci.* **110**, 2756–2762 (2008).
[27] Y. Abe, A. E. Tonelli, and P. J. Flory. *Macromolecules* **3**, 294–303 (1970).
[28] R. H. Boyd and S. M. Breitling. *Macromolecules* **5**, 279 (1972).
[29] U. W. Suter, S. Pucci, and P. Pino. *J. Am. Chem. Soc.* 97, 1018 (1975).
[30] U. W. Suter and P. J. Flory. *Macromolecules* **8**, 765 (1975).
[31] O. Schwarz. *Polymer Materials Handbook*. Natal Witness Press, Pietemaritzburg (1995).
[32] A. Valenza, F. P. La Mantia. Recycling of polymer waste: part II—Stress degraded polypropylene. *Poly. Degrad. Stab.* **20**, 63–73 (1988).
[33] V. Flaris and Z. H. Stachurski. *J Appl Polym Sci*, **45**, 1789–1798 (1992).
[34] P. Galli, L. Luciani, and G. Cecchin. *Die Angew. Makromol. Chem.* **94**, 63–89 (1981).
[35] F. Nakatsubo, and T. Takano, T. Kawada, and K. Murakami. In (Kennedy, J. F., Phillips, G. O., and Williams, P. A., eds) *Cellulose, Structural and Functional Aspects*. Ellis Horwood, Sussex, 201 (1989).
[36] S. Kobayashi, K. Kashiwa, T. Kawasaki, S. Shoda. *J Am Chem Soc.* **113** 3079 (1991).
[37] K. Endo, *Prog. Polym. Sci.* **27**, 2021–2054 (2002).
[38] W. H. Jr. Starnes, *J. Polym. Sci. Polym. Chem.* **43**, 2451–2467 (2005).
[39] J. E. Mark, *J. Chem. Phys.* **56**, 451 (1972).
[40] G. D. Smith, P. J. Ludovice, R. L. Jaffe, and D. Y. Yoon, *J. Phys. Chem.* **99**, 164 (1995).
[41] E. H. Gordon, F. G. Arthur, and G. S. *Walter, In Fire and Polymers: Hazards Identifications and Prevention*. (Nelson G. L. Ed.) American Chemical Society, Washington, DC, 12 (1990).
[42] G. Montaudo. *Polym. Degrad. Stab.* **33**, 229 (1991).
[43] J. L. Rogers, J. Vinyl. *Tech* **6**, 54 (1984).
[44] T. Jando, T. Stelzer, and F. Frakas, J. Electrostat, **23**, 117 (1989).

[45] E. P. Jr. Moore. *Polypropylene Handbook, Polymerization, Characterization, Properties, Processing, Applications*, Hanser, Munich (1996).
[46] N. L. Rodrigues, R. Jr. Antonio, W. Ormanji. *Tecnologia do PVC, Braskem.* Pro Editores (2002).
[47] D. Willoughby. *Plastic Piping Handbook.* McGraw-Hill, New York, (2002).
[48] J. K. Sears, J. R. Darby. *Technology of Plasticizers.* John Wiley & Sons, New York 1982.
[49] L. I. Nass. *In Encyclopedia of PVC*, (2nd ed.) Marcel Dekker, Inc, New York, 1 (1986).
[50] E. J. Wickson, *Handbook of PVC Formulation.* John Wiley & Sons, New York (1993).
[51] L. F. Thompson, C. G. Willson, and J. M. J. Frechet. *Materials for Microlithography: Radiation-Sensitive Polymers.* American Chemical Society, Washington, DC 266 (1984).
[52] M. S. Htoo. (ed.). *Microelectronic Polymers.* Marcel Dekker, New York, (1989).
[53] K. Matyjaszewski, T. P. Davis. *Handbook of Radical Polymerization.* John Wiley and Sons, Hoboken (2002).
[54] S. Mecking. *Angew. Chem. Int. Ed.* **43**, 1078–1085 (2004).
[55] R. A. Kudva. H. Keskkula, D. R. Paul. *Polymer* **41**, 225 (2000).
[56] Paul, D. R. Barlow, J. W. J *Macromol. Sci. Rev. Macromol. Chem.* **18**, 109 (1980).
[57] L. A. Utracki. *Polymer Blends and Alloys: Thermodynamics and Rheology.* Hanser, Munich, 1 (1989).
[58] D. G. Legrand, and J. T. Bendler, *Handbook of Polycarbonate Science and Technology*, Marcel Dekker, Inc., New York, (2000A).
[59] J. Roovers. in *Encyclopedia of Polymer Science and Engineering.* New York, (1985).
[60] D. G. Legrand, and J. T. Bendler, *Handbook of Polycarbonate Science and Technology.* Marcel Dekker, Inc., New York, (2000).
[61] M. J. Marks, S. Munjal, S. Namhata, D. C. Scott, and F. Bosscher, *J Polym Sci Polym Chem Ed.* **38**, 560 (1999).
[62] V. DeMaio, D. Dong, and A. Gupta, *ANTEC*, **799** (2000).
[63] D. Tsiourvas, E. Tsartolia, and A. Stassionopoulos. A new approach to reclaimed PET utilizations Blends of recycled PET suitable for extrusion blow-molding technology. *Adv. Polym. Technol.* **14**, 227–236 **(1995).**

[64] F. La Mantia. *Handbook of Plastics Recycling*. Rapra Technology Ltd, Shropshire, (2002).
[65] Whinfield. *Nature* **158**, 930 (1946).
[66] Hill and Walker. *J. Polymer Sci.* **3**, 619 (1948).
[67] Y. J. Jadhav. Kantor, S. W. *Polyesters, Thermoplastic*. In (Kroschwitz J. I. ed.). *High Performance Polymers and Composites*. John Wiley & Sons, New York 730–747 (1991).
[68] T. E. Attwood, P. C. Dawson, J. L. Freeman, L. R. J. Hoy. and J. B. Rose. P. A. Staniland. *Polymer* **22**, 1096 (1981).
[69] A. Jonas. R. Legras, *Polymer* **32**, 2691 (1991).
[70] D. P. Jones. D. C. Leach, D. R. Moore. *Polymer* **26**, 1385 (1985).
[71] J. Y. Chen, M. Chen, and S. C. Chao. *Macromol. Chem. Phys.* **199**, 1623 (1998).
[72] A. Jonas, and R. Legras. *Polymer* **32**, 2691 (1991).
[73] D. P. Jones, D. C. Leach, and D. R. Moore. *Polymer* **26**, 1385 (1985).
[74] J. Y. Chen, M. Chen, and S. C. Chao. *Macromol. Chem. Phys.* **199**, 1623 (1998).
[75] A. Novak, and E. Whalley. *Trans. Faraday Soc.* **55**, 1484 (1959).
[76] M. Mucha *Colloid Polym. Sci.* **162**, 103 (1972).
[77] Y. Yamashita, T. Asakura, M. Okada, and K. Ito. *Makromol. Chem.* **129**, 1 (1969).
[78] A. Abe and J. E. Mark. *J. Am. Chem. Soc.* **98**, 6468 (1976).
[79] P. J. Flory and J. E. Mark. *Makromol. Chem.* **75**, 11 (1964).
[80] G. D. Smith, R. L. Jaffe, and D. Y. Yoon. *J. Phys. Chem.* **98**, 9078 (1994).
[81] Martin, R. W. *The Chemistry of Phenolic Resins*. John Wiley and Sons, New York, 12 (1956).
[82] Martin, R. W. *The Chemistry of Phenolic Resins*. John Wiley and Sons, New York, 262 (1956).
[83] Widmer, G. *In Encyclopedia of Polymer Science and Technology*, (H. F. Mark ed.). John Wiley and Sons, New York, **2**, 54 (1965).
[84] W. G. Potter. *Epoxide Resins*. Springer, New York, (1970).
[85] C. A. May and G. Y. Tanka. *Epoxy Resin Chemistry and Technology*. Marcell Dekker, New York, (1973).
[86] M. J. Scudamore. *Fire Mater* **18**, 313 (1994).
[87] G. T. Egglestone, and D. M. Turley. *Fire Mater* **18**, 255 (1994).
[88] A. P. Mouritz, Z. Mathys. *Compos. Struct.* **47**, 643 (1999).
[89] B. K. Kandola, A. R. Horrocks, P. Myler, and D. Blair. *Composites Part A* **33**, 805 (2002).

[90] S. Hörold. *Polym. Degrad. Stab.* **64**, 427 (1999).
[91] P. Vanhoorne, P. Dubois, R. Jerome, and P. Teyssie. *Macromolecules* **25**, 37 (1992).
[92] F. Kawai. *Crit. Rev. Biotechnol.* **6**, 273 (1987); R. Keeler. *Res. Dev.* 52–57 (1991).
[93] J. E. Potta. et al. *Aspects of Degradation and Stabilization of Polymers*. Elsevier, Amsterdam, 671 (1978).
[94] R. Narayan. *Polymeric Materials from Agricultural Feedstocks, Polymers from Agricultural Coproducts*, (M. L. Fishman, R. B. Friedman and S. J. Huang, eds), ACS, Washington DC (1994).
[95] R. Narayan, and R. *Rationale, Drivers, and Technology Examples, Biobased & Biodegradable Polymer Materials*, (K. C. Khemmani and C. Scholz, eds), ACS, Washington DC (2006).
[96] P. Vanhoorne, P. Dubois, R. Jerome, and P. Teyssie *Macromolecules* **25**, 37 (1992).
[97] S. H. Hyon, W. I. Cha, Y. Ikada, M. Kita, Y. Ogura and Y. Honda. *Biomater. Sci. Polym. Ed.* **5**, 397 (1994).
[98] J. H. Junag, W. S. Bonner, Y. J. Ogawa, P. Vaconti and G. C. Weir. *Transplantation* **61**, 1557 (1976).
[99] D. H. Chen, J. C. Leu and T. C. Huang. *J. Chem. Technol. Biotechnol.* **4**, 351 (1994).
[100] C. L. McCormick, J. Bock, and D. N. Schulz. *Encyclopedia of Polymer Science and Engineering,* John Wiley, New York, NY, **17**, 730 (1989).
[101] Pitha. *J. Adv. Polym. Sci.,* **50**, 1, (1983).
[102] J. F. Kennedy, C. A. White, and E. in Haslam. *Comprehensive Organic Chemistry,* Vol. 5, Pergamon Press, Oxford, UK.
[103] H. F. Zhou, Q. H. Fan, W. J. Tang, L. J. Xu, Y. M. He, G. J. Deng, L. W. Zhao, L. Q. Gu, and A. S. C. Chan. *Adv. Synth. Catal.*, **348**, 2172–2182 (2006).
[104] M. S. Thompson, T. P. Vadala, M. L. Vadala, Y. Lin, and J. S. Riffle. *Polymer*, **49**, 345–373 (2008).
[105] R. Bhardwaj, and A. K. Mohanty. *Biomacromolecules*, **8**, 2476–2484. (2007).
[106] S. Lee, and J. W. Lee. *Korea Aust. Rheol. J.*, **17**, 71–77, (2005).
[107] Y. Doi, K. Sano, H. Sawada, H. Tanaka, Y. Tokiwa, K. Fukuda, H. Matsumura, and Y. Yoshida. In (Doi, Y., ed.) *Biodegradable Plastic Handbook*, Biodegradable Polymer Institute Publisher (1995).
[108] R. G. Sinclair. *JMS Pure Appl. Chem.*, **A33,** 585 (1996).

[109] S. R. Bhattarai, N. Bhattarai, H. K. Yi, P. H. Hwang, D. I. Cha, and H. Y. Kim. *Biomaterials*, **25**, 2595 (2004).
[110] O. M. Bostman, and J. Bone. *Joint Surg.*, **73A**, 148 (1991).
[111] J. W. Leenslag, A. J. Pennings, R. R. M. Bos, F. R. Rozema, and J. Boering. *Biomaterials*, **8**, 70 (1987).
[112] M. Todo, S. D. Park, T. Takayama, and K. Arakawa. *Eng. Fract. Mech.*, **74**, 1872 (2007).
[113] J. W. Park, and S. S. Im. *Polym. Eng. Sci.*, **40**, 2539 (2000).
[114] T. Ke, and X. Sun. *J. Appl. Polym. Sci.*, **81**, 3069 (2001).
[115] H. X. Sun, and P. Seib. *J. Appl. Polym. Sci.*, **82**, 1761 (2001).
[116] H. Wang, X. Sun, and P. Seib. *J. Appl. Polym. Sci.*, **84**, 1257 (2002).
[117] T. Ke, and X. Sun. *J. Appl. Polym. Sci.*, **88**, 2947 (2003).
[118] H. Wang, X. Sun, and P. Seib. *J. Appl. Polym. Sci.*, **90**, 3683 (2003).
[119] J. F. Zhang, and X. Sun. *J. Appl. Polym. Sci.*, **94**, 1697 (2004).
[120] J. F. Zhang, and X. Sun. *Biomacromolecules*, **5**, 1446 (2004).
[121] S. B. Shin, G. S. Jo, K. S. Kang, T. J. Lee, and B. S. Kim. *Macro mol. Res.*, **15**, 291 (2007).
[122] Y. J. Kim, Y. M. Lee, and H. M. Lee. *Korean J. Chem. Eng.*, **11**, 172 (1994).
[123] H. S. Byhn and H. Y. Lee. *Korean J. Chem. Eng.*, **23**, 1003 (2006).
[124] A. Schindler, R. Jeffcoat, G. L. Kimmel, C. G. Pitt, M. E. Wall, and R. Zweidinger. In (Pearce, E. M., and Schaefgen J. R., eds), *Contemporary Topics in Polymer Science*, Plenum, New York, **2**, 251–289 (1977).
[125] C. G. Pitt, T. A. Marks, A. Schindler. *"Biodegradable Drug Delivery Systems Based on Aliphatic Polyesters: Application to Contraceptives and Narcotic Antagonists", in Controlled Release of Bioactive Materials*, (Baker, R., ed.) Academic Press, New York, 19–43 (1980).
[126] B. Buntner, M. Nowak, M. Bero, P. Dobrzynski, J J. Kasperczyk. *Bioact. Compat. Polym.*, **11**, 110–121 (1996).
[127] C. G. Pitt, and A. Schindler. *"Biodegradable Of Polymers", in Controlled Drug Delivery*, (Bruck S. D., ed.), Boca Raton, EL, CRC Press, 55–80 (1983).
[128] C. J. Goodwin, M. Braden, S. Downes, and N. J. Marshall. *J. Biomed. Mater. Res.*, **40**, 204–213 (1998).
[129] A. J. Dombs, S. Amselem, and M. Maniar. In (Dumitriu, S., ed.), *Polymeric Biomaterials*, Dekker, New York, **I**, 399 (1994).

[130] S. Li and M. Vert. In (Scott, G. and Gilead D., eds), *Degradable Polymers*, Chapman and Hall, London, **43** (1995).
[131] E. E. Schmitt, and R. A. Polistina. *Surgical Sutures, US Patient 3, 297 033*, January 10, (1967).
[132] A. M. Reed, and D. K. Gilding. *Polymer,* **22**, 494 (1981).
[133] C. C. Chu. *J. Appl. Polym. Sci.,* **26,** 1727 (1981).
[134] M. Vert, S. M. Li, G. Spenlehauer, and P. Guerin. *J. Mater Sci. Mater. Med.*, **3,** 432 (1992).

4

Polymer Chemistry

Polymer chemistry is a part of chemistry (Figure 4.1). It deals from atoms to chemical reactions to final products. Polymer chemistry exists in a close relationship with product development (Figure 4.1). Polymers, made from monomers, are much more complex, and their origin is the basic problem in understanding the origin of life.

In polymer chemistry, innovations are continuously taking place. There are in-depth aspects of chemical reactions in order to form polymers. Due to their properties, excellent chemical resistance, good processing ability and assembly simplification due to potential for part consolidation polymers are made as acceptable materials. Polymer owes its success to opportunities of new synthesis which arise from chemistry. Dramatic increases in polymers marked as the history of the development in the living conditions (Figure 4.2). Polymer has had a great influence on civilization than other technological products.

Polymer chemistry is concerned with covalent bonds. Polymers can exist as chains or rings. Structure levels are defined in a solid polymer as:

- Constitution-based on atomic structure,
- Configuration-based on primary bonds
- Conformation-based spatial arrangement of the complete chain, even in the case of semi-crystalline polymers
- Morphology on molecular structure.

The relationship between constitution, configuration and attainable conformations of individual polymer and the elastic behaviour of polymer vary substantially with the change in temperature.

They are as [1]:

- Constitution i.e. atomic structure – the polymer is made up of either pure hydrocarbons or substituted hydrocarbons or with different elements such as sulphur, phosphorus, nitrogen, etc.

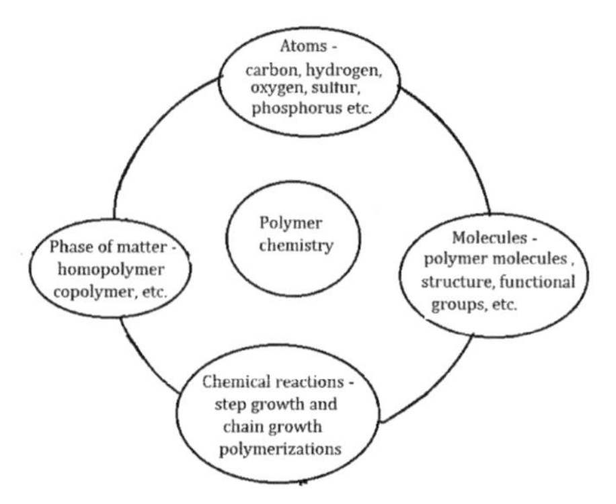

Figure 4.1 Polymer chemistry.

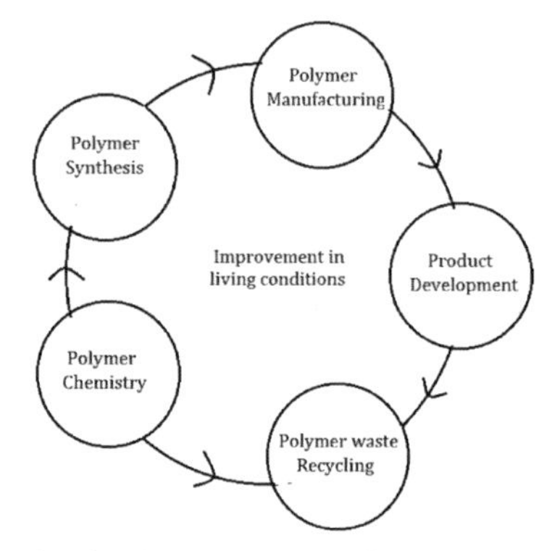

Figure 4.2 Interrelationship of polymer chemistry in the improvement of living conditions.

- Configuration i.e. primary bonds – the backbone chains form by covalent bonds. The results of configuration based on bond length. Carbon – carbon bonds result in a simple zigzag planar configuration in polyethylene. In polymers, weak intermolecular interactions, such as van der Waals and strong hydrogen bonds exist between neighbouring macromolecules, exception network polymers [2]. In polymer, the regularity of backbone determines the molecular structure and its control such as stereochemistry. Molecular orientation affects the mechanical properties of the material.

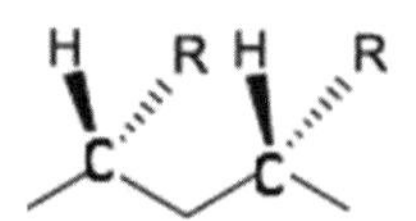

- Conformation i.e. spatial arrangement – conformation results from the steric hindrance in the spatial packing of segments lead to a helix configuration of the backbone. The entropy ΔS of conformation characterize the number of accessible conformation at a given external conditions [3].

4.1 Types of Polymers

Polymers form from the polymerization of monomers. They are structurally long chains with repeating structural units. To form cyclic or saturated structure from univalent groups, it may be joined conceivably. They are commonly two major types named as homo-polymers and copolymers. Polymer ends with a functional group such as $CH_3 - CH_2 - [CH_2\ CH_2] - CH = CH_2$ are called telechelic polymer. Reactive oligomer contains end groups, which are capable to undergo polymerization.

The polymer chain may typical not only differ with respect to their chain length (molecular weight), but also in their end groups, architecture and chemical composition. Simple polymer with linear or branched has different tacticities and functionalities. In copolymers, the chemical composition can vary from chain to chain, but also along each chain.

4.1.1 Homopolymers

Homopolymers are from a single kind of monomer. They consist of a single repeating unit, Majority of polymers are homopolymers with the same monomer repetition such as terminal alkenes produced, such as in polyethylene. There are polymers made from two different monomers whose structural units form the repeating unit such as the formation of polyamide from diamine and a diacid.

$$\sim\sim M - M - M - M - M - M - M - M \sim\sim$$

Homopolymer

Where $M = [-CH_2 - CH_2 -]$, $[-CO-(CH_2)_4-CO-NH-(CH_2)_6-NH-]$, etc.

4.1.2 Copolymers

Copolymers are polymerization with more than one kind of monomer, which is made up of two or more repeating monomer units. Copolymer includes chain topologies of statistical and also periodic, gradient and segmented such as blocks and grafts. However, the chain architecture varies which include

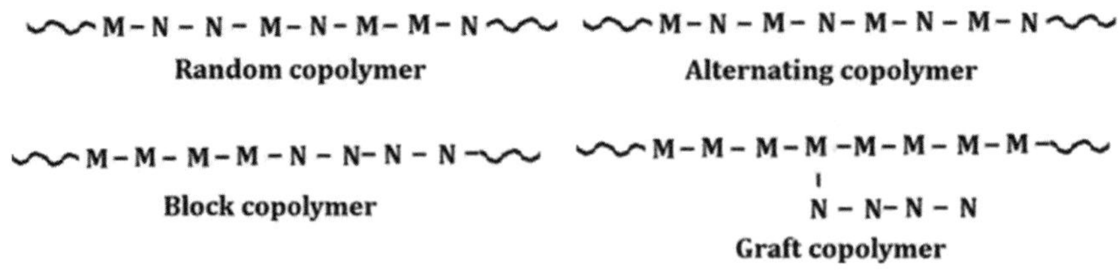

〰M - N - N - M - N - M - M - N〰 〰M - N - M - N - M - N - M - N〰
Random copolymer Alternating copolymer

〰M - M - M - M - N - N - N - N〰 〰M - M - M - M - M - M - M - M〰
N - N - N - N
Block copolymer Graft copolymer

Figure 4.3 List of copolymers.

comb, multi-arm stars, and dendrimers or even growth from functionalized surfaces. The possibilities for compositional modification of different copolymers are almost limitless.

For copolymers, one type of arrangement may involve the number of M and N units which are present in the sequences and, also, the distribution of these sequences. The random placements of M and N follow from the well-known copolymer equation [4]. They are classified as random, alternating, block and graft copolymer (Figure 4.3).

Efficacy generally refers to the effectiveness of the copolymer to situate itself at the interface and reduce the interfacial tension and particle-particle coalescence [5–9]. The copolymer molecular weight has little influence on the plateau value and phase size [10–11]. An increase in the copolymer molecular weight also increases the tendency of copolymer micelle formation [12–16].

4.1.2.1 Random copolymers

The composition along a single chain is approximately the same in random copolymers. Absence of regularity would make a copolymer as random copolymer. Free radical polymerization is used to prepare random polymers. A chain formed at low conversion may have different composition based on copolymerization parameters and monomer feed ratio at a later stage of polymerization. The majority of commercially available polyolefin based on either homo-polymer or random copolymer architectures. Some cationic drugs show antitumor activity such as random copolymer of low molecular weight consists of ethylene and modified maleic acid units [17–18].

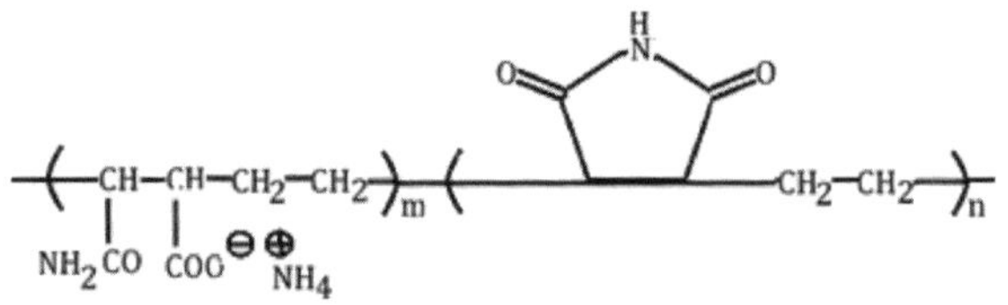

$$-\left(CH(CONH_2)-CH(COO^{\ominus}\,NH_4^{\oplus})-CH_2-CH_2\right)_m-\left(\text{succinimide}-CH_2-CH_2\right)_n-$$

4.1.2.2 Block copolymers

Block copolymers contains two different monomers. The synthesis of block copolymers is industrially done by anionic polymerization. In addition to the random-sequence and an alternating one, sometimes called ordered-sequence, there are also block copolymers. These are copolymers made up of blocks of individual polymers joined by covalent bonds. An example can be a block copolymer of styrene and isoprene:

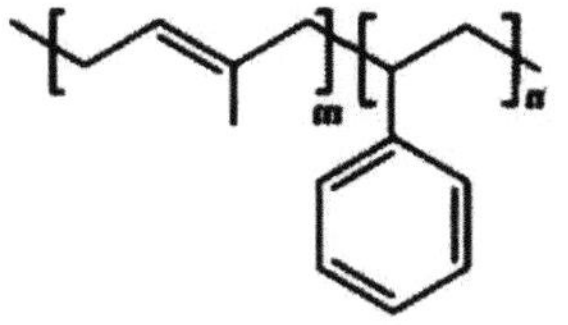

Polystyrene block – polyisoprene block

4.1.2.3 Alternating copolymers

There should be regularity in the repetition of the structural units and repetition alternate, the copolymer is called alternating copolymer. An example of an alternating copolymer can be a copolymer of styrene with maleic anhydride:

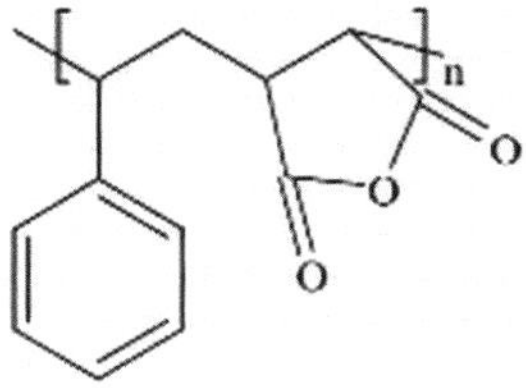

4.1.2.4 Graft copolymer

An ideal graft or an exact graft copolymer is one in which all of the above parameters perfectly controlled [19–22]. Graft copolymers consist of two or more numbers of branch chains connected to the backbone chain. They are multiphase materials exhibit unique and interesting morphologies due to the backbone and graft chains are thermodynamically incompatible [23–24].

Graft copolymer is defined by:

- Molecular weight of the backbone chain
- Molecular weight of the graft chain
- Distance or molecular weight between the branch points
- Number of graft chains along the backbone chain.

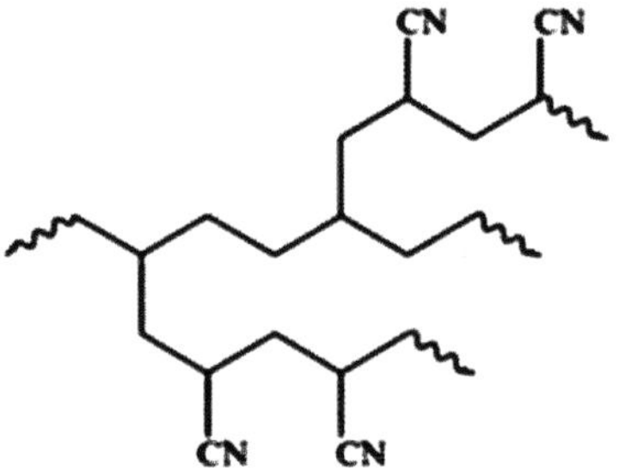

The amount of homopolymer and the copolymer as well as the number and length of the grafts have to be determined. The length of the macromonomer decides the length of the grafts. The copolymer possesses backbone composed of one individual polymer and branches from another one commonly known as graft copolymer. The backbone and branches graft during polymerization. Many graft polymers form from a different polymer backbone by polymerizing the branch polymers. Polyacrylamide on polyethylene is an example of graft copolymer.

4.2 Polymer Morphology

Polymer morphology refers to the macromolecular structure level or order of the long chain. Molecular structure of polymer is a complex structure and polymerization process leads to the formation of long molecule. The molecular structure controls the macromolecular formation. In polymer, the conformations and morphology control the mechanical properties of glassy and semi-crystalline polymers respectively [25]. Polymer chains interact during solidification/crystallization which occurs upon cooling. It creates either long range ordered crystalline or disorder amorphous phases. Constitution and configuration govern the melting point of the crystalline phase and the glass transition temperature of the intervening amorphous regions, largely through controlling the main chain stiffness.

The term "structure" covers a broad range of possible interpretations. Structural levels in a solid polymer are [26] determined by chemical structure. Monomer can be of single, double or triple bonds. Monomers are low molar mass molecules. Polymer structures complicate the relationships between various structural parameters and physical properties. The structure variables affect the mechanical properties of polymer. It is essential to engineering and tailoring polymer in order to control the mechanical properties that need to decide products, applications, design and processing technique.

Small molecules such as structural unit have low molecular mass. Polymer structure covers a broad range of interpretation. Polymer structures affect the flow and morphology, which result in different physical properties. Polymers have chemical structure of long chain molecules. Polymer builds structurally from a repeating radical or structural unit. The general formula of one of the polymer is

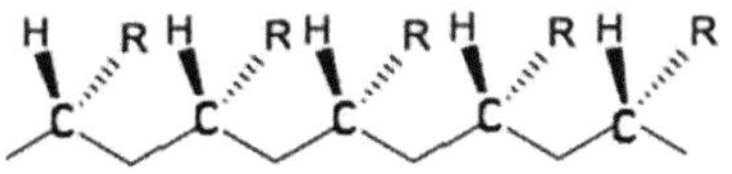

Polymer 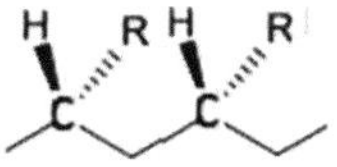 **Structural unit**

Where R = H, Cl, C_6H_5, etc.

Polymer structure indicates flexibility and rigid nature. The hydrogen-bonding present in the polymer such as nylon, etc., enhances rigidity and makes solvent resistance. The long backbone chain provides mechanical properties. The structural features affect the accessibility of various chain conformations and thus are pivotal for a control of the mechanical behaviour of polymers and for the morphology of the polymers.

4.3 Stereochemistry

In polymer, the regularity of the molecular backbone determines its control such as stereochemistry. The polymer structure distinguishes itself from atomic materials. The constitution and configuration are directly reflecting the mechanical properties. Polymer having identical constitutional repeating units can nevertheless differ because of isomerism. Linear, branched, and cross-linked polymers of the same monomer are considered as structural isomers. The number of conformation and morphology control the mechanical properties. Stereoisomerism occurs from differences in configuration of asymmetric carbon atoms in the chain. Properties of polymers have associated with their structure with the configuration of units in the polymeric molecules.

Structural isomerism formation is due to branches in homopolymer and variation in the monomer distribution in copolymers. Structural variations in polymer occur, they are

1. Sequence isomerism
2. Stereoisomerism
3. Geometric isomerism

4.3.1 Sequence Isomerism

Sequence isomerism arises from the variations in orientation of asymmetric monomer units. Most monomers used chain growth polymerizations are non-symmetrical results in three possible sequences of the monomer units in the chain. If the chain is composed of substituted monomeric units linked together on a regular basis, "head to tail" the spatial location of the substituents determines the tacticity of the polymer. These include head – to – tail, tail – to – tail and head – to – head structures termed as sequence isomerism.

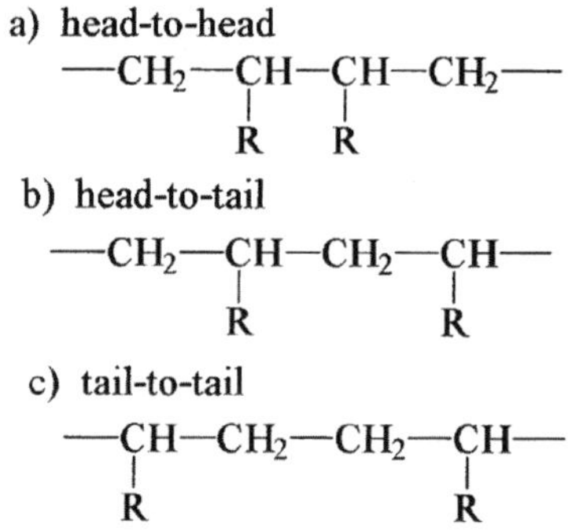

a) head-to-head

$$-CH_2-\underset{R}{\underset{|}{CH}}-\underset{R}{\underset{|}{CH}}-CH_2-$$

b) head-to-tail

$$-CH_2-\underset{R}{\underset{|}{CH}}-CH_2-\underset{R}{\underset{|}{CH}}-$$

c) tail-to-tail

$$-\underset{R}{\underset{|}{CH}}-CH_2-CH_2-\underset{R}{\underset{|}{CH}}-$$

Sequence isomerism in vinyl polymers

4.3.2 Stereoisomerism

Linear polymer can be amorphous i.e., noncrystalline or semicrystalline. Polymers such as polypropylene (Figure 4.4) with asymmetric carbon atoms generally can be found in the amorphous or semicrystalline states depending upon their configuration. Polymers without asymmetric carbon atoms generally crystallize. Chain regularity is an attribute of high levels of structural hierarchy. Stereoisomerism affects the morphology and mechanical properties of semicrystalline polymers.

Stereoisomerism is the structural variation in the position of small side groups along with the long chain polymer arising from the presence of asymmetric centres formed during polymerization. Asymmetric carbon in a polypropylene monomer is a typical example. A regular structure of atactic PP can be produced in a single polymerization reaction.

(i) isotactic (ii) syndiotactic or (iii) atactic configuration

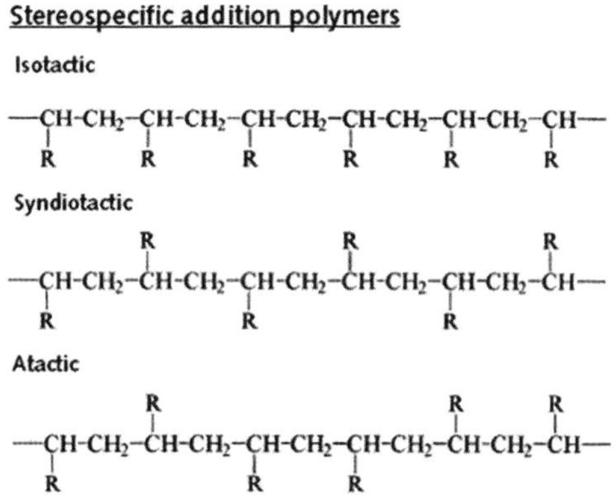

Stereospecific addition polymers

Isotactic

—CH-CH$_2$-CH-CH$_2$-CH-CH$_2$-CH-CH$_2$-CH-CH$_2$-CH—
R R R R R R

Syndiotactic

R R R
—CH-CH$_2$-CH-CH$_2$-CH-CH$_2$-CH-CH$_2$-CH-CH$_2$-CH—
R R R

Atactic

R R R
—CH-CH$_2$-CH-CH$_2$-CH-CH$_2$-CH-CH$_2$-CH-CH$_2$-CH—
R R R

Figure 4.4 Polymer having an asymmetric carbon atom.

In polypropylene, three different configurations found. One with the side chain group – CH_3 completely above or below the plane of the chain called isotactic. The side groups may be found alternatively above and below in order of the plane yielding a configuration called syndiotactic. The side groups of the chain configuration above and below are randomly placed is called as atactic. In isotactic and syndiotactic configuration, the regular arrangement of the chains leads to semicrystalline polymers. Atactic configuration leads to amorphous polymers.

4.3.3 Geometric Isomerism

Structural isomerism finds in polymeric conjugated dienes. The structural features affect the accessibility of the mechanical behaviour of polymers and for the morphology of semicrystalline polymers. Addition of monomer to the chain end can occur in 1,2 and in 1,4 position. In nonsymmetric dienes, 3,4 addition is further possible. Polymers with double bonds within their main chain such as poly (1,4) isoprene arises geometric isomerism in diene polymers as cis and trans structures. 1,4 cis or trans polydienes are very important for a degree of crystallinity and the physicochemical properties.

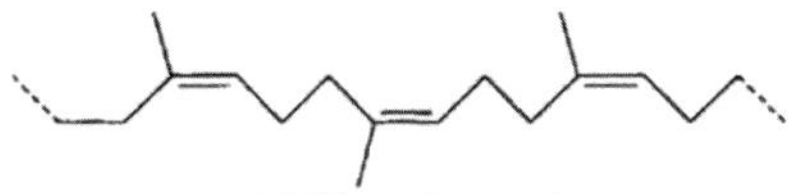

Poly(cis-1,4-isoprene)

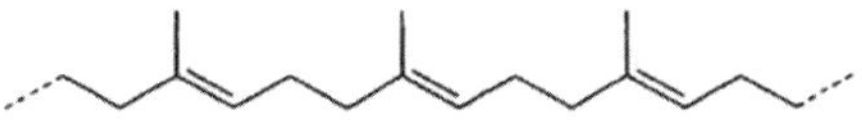

Poly(trans-1,4-isoprene)

4.4 Polymer Architecture

Polymer forms a long chain apart from other chemical species. The long chain molecule gives unique characteristic properties. Therefore, in the chemical engineering, polymerization will have more prominent place. Polymer chain can be linear, branched or cross-linked. Polymer architecture depicts by their most common constitutional repeating units, i.e. idealized structures. The groups at the end of their chains illustrate in part because they are often unknown. Their structure does not influence most of the polymer properties.

4.4.1 Linear Polymers

Linear polymers or linear chains exhibit simplest structures. They possess monomer groups at both ends. It forms due to their dimensional connectivity or their unbranched structures. Linear polymers such as acrylics, nylons, etc., and branched polymers such as polyethylene are thermally soften and on cooling, becomes rigid. The chain is open or closed with linear structure will be called as linear polymers.

Linear polymers have properties which include:

- Higher melting point;
- Low solubility;
- Greater hardness;
- Stiffness.

The unusual elastic and flow properties of systems containing linear polymeric molecules arise from three factors: (1) the length of the polymer molecules, (2) the flexibility of the molecular chains, and (3) the interactions of the segments of a polymer molecule with other segments of the same and other polymer molecules [27]. Commercially important linear polymers are Polyester, Polyurethane, Polycarbonate, Polyamide, Polyethers, Polyurea, Polysulphide and Polysulphone.

The figure represents the schematic representation of linear and branched polymers in Figure 4.5.

In linear polymers, the number of conformations and the morphology control the mechanical properties of glassy and semicrystalline polymers.

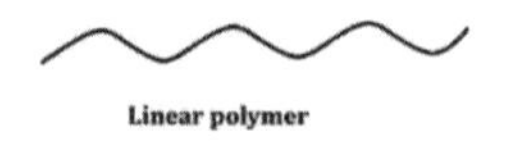

Figure 4.5 Linear polymers.

4.4.2 Branched Polymers

Irregular structures may create by chain-transfer reactions as polymer molecules. The reactions lead to branched polymers. In polyethylene, the polymerization reaction products leads to polymers as illustrated in Figure 4.6 with branches. The melting temperature and density of branched polymers are lower than linear counterparts. The branched polymers exhibit complex rheological behaviour during extrusion, including the appearance of a succession of defects in the extrudates that are limiting factors in processing [28]. High-pressure polyethylene industrially is obtained at high pressure and temperature and is considered a branched polymer. The schematic representation of branched polymer is as in Figure 4.6.

4.4.3 Network Polymers

Network polymers (Figure 4.7) are of macromolecule(s) consisting of number of conjoined molecules. Each conjoined molecules have at least three subchains in common with neighbouring molecules. Each constitutional unit connects to the other constitutional unit and to the macroscopic phase boundary by many permanent paths. The average number of intervening bonds increasing with the number of paths. In network polymers, some pendant linear and branched chains with free ends occur. It is prepared by a crosslinking of an original linear or branched polymer. A swollen network polymer is a gel with a swollen micro work called micro gel.

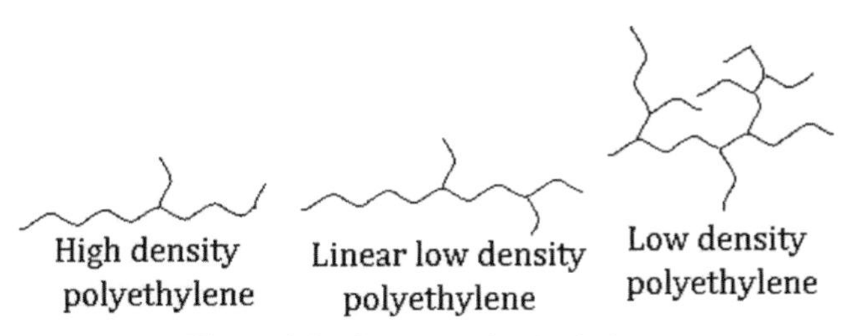

Figure 4.6 Structure of polyethylene.

Figure 4.7 Network polymers.

4.5 Polymer Properties

Polymer synthesis in which the monomer units arranged in a specific manner, thereby the chains control the properties of polymers. Polymers differ significantly from monomers in physical properties. Molecular structure controls the physical properties of polymer [29]. Copolymers are different in nature even though it has the same number of monomers. The number and nature of structural units determine the molecular nature. The symmetry and asymmetry of chemical structure affect the physical properties of polymers (Figure 4.8).

A polymer has an identical composition. However, the individual chain molecules differ in their structure, configuration, conformation, as well as in their molecular size. There is a mixture of molecules of different size in terms of molecular weight distribution. Most of the physical measurements such as molecular weight only give average values. Linear polymers have high melting with low solubility. They have great hardness and stiffness. However, the polyethylene polymer is brittle at ordinary temperatures. Polymers can expand or contract easily upon application of an external force. Polymers with numerous cross-links pull them back into their original shape by the removal of stress. Internal energy is independent of the extension of the random coil in a perfect polymer.

Polymers are strongly viscoelastic in nature and great sensitivity of physical properties based on external effects such as temperature pressure, time and deformation [30]. A polymer's melting viscosity depends very sensitively on the molecular weight, molecular weight distribution and on branching.

4.5.1 Microscopic Properties

Polymer chemistry on the microscopic scale continues to grow in importance and to impact key applications in the fields of materials science and technology. The development of advanced polymeric materials for

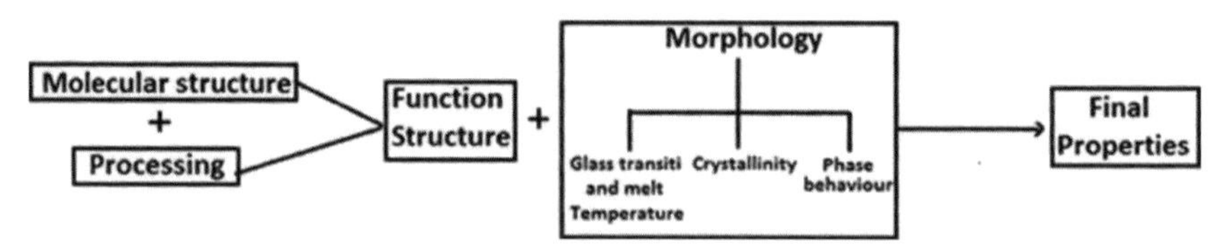

Figure 4.8 Polymer structure and processing provides its function, structure in addition with morphology and together provides final properties.

applications requires detailed information about the physical and chemical properties. Polymer chain orientation plays a major role in mechanical properties [31–33].

Microscopic parameters affect the mechanical response of the polymer. They are:

- Chemical structures of long chain molecules
- Molecular weight and molecular weight distributions
- Backbone chain stiffness
- Molecular shape
- Molecular orientation
- Intermolecular interactions.

Polymers are also polydispersed in nature due to their molar mass, chemical composition, functionality and molecular architecture [34, 35]. Polymers are classified as either natural or synthetic. Natural polymers found in nature as resin, gum. Synthetic polymers are produced from manufactured chemicals. Out of the synthetic polymers, plastics classified as thermoplastics or thermosets. Today plastics have become more common in industrial applications.

4.5.2 Crystallinity

Crystalline structures initially produced, seldom represent even close approaches to states of thermodynamic equilibrium [36]. The polymer crystallinity is the extent to which the molecules have a straight or a branched structure. Constitution and configuration govern the melting point of the crystalline phase. The long chain molecule to crystallize upon the chain regularity is conditional at all to permit packing in parallel orientation of reasonably well-ordered lattices. The ability Polymer encounters symmetry i.e. translational order in crystalline phase. Polymers such as PMMA, SAN, ABS, PS, PVA, PVAc and others are commercially important polymers with absence of symmetry [37].

Semicrystalline polymers are less sensitive to temperature, more environmentally stable, more rigid and less permeable to low molar mass agents compared to the majority of polymers [38]. Increase in crystallinity occurs with molecular weight. The structural effect on the symmetry of the crystallinity and asymmetry in side effects, the total crystallinity remains in the polymer [39]. Phase transformation shows a wide variety of microstructures in polymers.

4.5.3 Gel Effect

Free-radical polymerizations are typically highly exothermic. An increase in temperature will increase the polymerization rate; generate heat dissipation and a tendency to develop localized "hot spots". Near the end of polymerization, the viscosity is very high and difficult to control the rate as the heat is "trapped" inside. It leads to the auto-acceleration process (Trommsdroff or gel effect) in which the propagation rate is very higher than that of termination rate. This method is seldom used in commercial manufacture. The gel effect is influenced by two factors depending on the solution viscosity:

- Macro-radical trapping effect
- Macro-radical occlusion effect

The auto-acceleration (or gel effect, Trommsdorff effect) has been known for a long time in polymerization. This auto-acceleration, called the Norris-Trommsdorff effect, causes the formation of unusually high molecular weight polymers [40–43]. It is assumed that the transfer coefficients are ratios of rate constants for the transfer reactions concerned to that for the propagation reaction, are independent of conversion. Molecular weights averages or intrinsic viscosity on conversion, this assumption made implicitly. The rate constants for propagation and termination vary with the conversion, especially at low temperatures. The termination reaction involves two polymeric species more affected by conversion than the propagation reactions that only involve one, as with other monomers. This is the reason for the well-known "gel effect".

4.5.4 Intermolecular Forces

The intermolecular forces in polymers can affect dipoles in monomer units. Polymers containing non-hydrocarbon groups can form hydrogen bonds between adjacent chains. It can result in the high melting point of the polymers. It can have dipole-dipole interactions between the non-hydrocarbon functions. However, these are not as strong as hydrogen bonding and the melting points of such polymers will be lower than hydrogen-bonded polymers. However, the dipole-bonded polymers will have greater flexibility. In a true hydrocarbon polymer, such as polyethylene, the situation is different.

Polymer properties play a large part with the attractive forces. Polymer chains are lengthy, and inter-chain forces simplify far beyond the attractions between conventional molecules. Different side groups or chains on the polymer can tend the polymer to have ionic bonding or hydrogen bonding between its own chains. These stronger forces typically result in high physical properties such as higher crystalline melting points.

4.5.5 Dipole Moment

The attractive forces between polymer chains arise from weak van der Waal's forces. Figure of molecules shows often as surrounded by a cloud of negative electrons. As two polymer chains approach, their electron clouds repel one another. This has the effect of lowering the electron density on one side of a polymer chain, creating a slight positive dipole on this side. The charge is enough to attract the second polymer chain. However, since van der Waal's forces are weak, polymer such as polyethylene can have a lowering of melting temperature compared to other polymers [44–46].

Degree of conversion during polymerization increases some structural change occurs [47]. The main structural factors affect the physical properties are stereoregularity, molecular mass, and polydispersity. Many of the properties put their influence on crystallinity. Tailoring polymer properties in the polymerization stage requires profound insight into the relationships between catalyst, chain microstructure, crystallization properties, and final product properties [48].

4.5.6 Surface Behaviour

Hydrophobic side chains of acrylate polymers exhibit surface freezing which is an exception in nature. Surface freezing is an aspect of almost all the materials exhibit the opposite behavior during surface melting [49]. The monolayer of ordered side chains exist on the surface above the bulk melting temperature [50–52]. Linear alkanes and alkane analogues are the only other materials that show to exhibit surface freezing [53–54].

4.5.7 Solution Properties

Solution properties, such as solubility, viscosity, and phase behaviour, are highly dependent on the macrostructure of the chain and the chemical microstructure of the repeat units. The presence of acidic or basic functionality causes pH, electrolyte, and temperature-dependent behaviour. The size of single polymer chain is dependent on its molecular weight and morphology.

The molecular chain can be extended in a very dilute solution with a good solvent is used to dissolve the polymer. However, the single polymer chain is in coil form due to the interactions between solvent and polymer in solution.

These properties are largely due to hydrogen bonding and the intramolecular and/or intermolecular association of hydrophobic groups. The hydrophobic association provide significantly to modify the rheology of a system.

In dilute solution, polymer macromolecules of long chain separate due to the flexibility and stochastic features involvement. Monomers are distant from one another along the chain backbone. In the space, the monomers may get to be close to each other. Such monomer pairs can then chemically cross-linked by reaction. During crystallization, the concentration of polymer in solution measured as a function of crystallization temperature.

4.5.8 Thermodynamics

With a possible equilibrium, in the polymerization reaction certain concentrations of the monomer remain has reached equilibrium. Thus, polymerizations are analogous in several ways to the usual chemical reactions that can only proceed to high yield if the equilibrium between reactants and product(s) favour the product(s). Moreover, a suitable mechanism must be available which will permit the reaction to proceed. In the case of polymerization, a large number of monomer units must be involved in a propagation step to generate a polymer. This reminds that the Gibbs equation must yield a negative free energy change for the propagation reaction if high molecular weight to be achieved. The criteria of a ceiling temperature, is that requires the standard free energy change associated with any polymerization.

The free energy change due to polymerization is equal to the difference of the Gibbs free energies of the polymer segment and the monomer molecule:

$$\Delta G_p = G_{pol} - G_{mon}$$

$$\Delta G_p = (H_{pol} - TS_{pol}) - (H_{mon} - TS_{mon})$$

$$\Delta G_p = \Delta H_p - T\Delta S_p$$

Where ΔG_p = difference in Gibbs free energy of polymer, G_{pol} and G_{mon} are free energy of polymer and monomer respectively. H_{pol} and H_{mon} are enthalpy of polymer and monomer respectively. T is the temperature. S_{pol} and S_{mon} are entropy of polymer and monomer respectively. ΔH_p and ΔS_p are the change in enthalpy of polymer and entropy of polymer respectively. Both enthalpy and entropy of monomer (initial state) and polymer segment (final state) depend on the state of association. Spontaneous polymerization requires $\Delta G_p < 0$. Ionic polymerizations are commonly carried out in solution, thus ΔG_p depends on monomer and polymer concentrations (activities).

The polymer is composed entirely of repeating units of one kind, while real linear macromolecules terminated at both ends with structurally different units: end groups. For high-molecular weight polymers, the contribution of these end groups is so small that to be neglected. The large majority of polymers described are also uniform in structure. Thus, from the thermodynamic point of view a polymerization sufficiently well-described by a single propagation reaction.

This change made up of enthalpy and entropy contributions, which together with reaction temperatures define the magnitude and sign of the free energy. The chemical structure of the monomer affects the free energy of the polymerization in a number of ways. These include monomer, the presence of functional groups on the monomer, the geometrical or stereo chemical chain isomerism and solid-state morphology i.e. crystallinity that might be observed in the resulting polymer. Thermodynamics of polymerization does not formally differ from that of any other chemical reaction if by 'polymerization' the propagation step understands.

Thus, the total change of the free energy of conversion of one mole of monomer to polymer is:

$$\Delta G_p = \Delta G_p^o - RT \ln [M]_o + \frac{[M]_e}{[M]_o - [M]_e} RT \ln \frac{[M]_e}{[M]_o} + \frac{1}{n} RT \ln \frac{[M]_o - [M]_e}{n}$$

Where ΔG_p^o corresponds to the standard conditions

ΔG_p^o = change in Gibbs free energy of one mole of monomer

$[M]_o$ = molecular weight of one mole of monomer

$[M]_e$ = molecular weight at equilibrium

Usually high molecular weight polymer is formed (n >>1) and the last term of the right hand side of equation can be neglected. When the monomer equilibrium concentration is much lower than the initial concentration of monomer then equation is reduced to the well-known equation.

$$\Delta G_p = \Delta G_p^o - RT \ln [M]_o$$

For initial concentration of monomer close to the equilibrium, i.e. when the overall change of the free energy is close to zero

$$\Delta G_p^o = RT \ln [M]_o$$

and the equilibrium constant of polymerization becomes:

$$K_e = \exp(-\Delta G_p^o / RT) = [M]_o^{-1} = [M]_e^{-1}$$

Where K_e is the equilibrium constant

Equation indicates that if the initial concentration of monomer is high polymerization may still may be thermodynamically feasible even if ΔH_p° which is change in enthalpy of one mole of monomer has a small positive value.

The standard free energy change is given by a change of enthalpy and entropy:

$$\Delta G_p^o = \Delta H_p - T \Delta S_p^o$$

Thus, four general cases are possible depending whether ΔH_p° and ΔS_p° are positive or negetive which change in enthalpy and entropy of one mole of monomer resepctively:

$$\Delta H_p^o < 0 \quad \text{and} \quad \Delta S_p^o < 0$$

For the large majority of heterocyclic monomers polymerization is exothermic due to the release of the ring strain. The entropy of the system **($\Delta S_p^o < 0$)** usually also decreases due to the loss of translational entropy. Thus, **$\Delta H_p^o < 0$** and **$\Delta S_p^o < 0$** for a system with polymerization is possible only below a **$T_c^o = \Delta H_p^o / \Delta S_p^o$** certain level are called ceiling temperature. This temperature is highest for bulk polymerization and decreases with the final monomer concentration.

$$T_c = \Delta H_p^o (\Delta S_p^o + RT \ln [M]_o)$$

ΔS_p^o and R ln **$[M]_o$** have opposite signs for **$[M]_o > 1$**. Thus, with increasing **$[M]_o$** the overall value of the denominator decreases and **T_c** increases.

The ceiling temperature can be observed only for systems in which has **ΔH_p^o** a relatively small negetive valuse (> – 40KJ. Mol-1).

For higher negetive valuses of $\mathbf{\Delta H_p^o}$ for most vinyl polymerization $\mathbf{T_c}$ is well-above the decomposition temperature of monomer and polymer.

The polymerization of the polymer – monomer equilibrium monomer concentration $\mathbf{[M]_o}$ is given by

$$\ln [M]_e = \Delta H_p^o / RT - \Delta S_p^o / R$$

The position of the equilibrium depends mostly $\mathbf{\Delta H_p}$ on at least at moderate temperatures. The positive charge of entropy of polymerization is due to the fact, that the rotational entropy of the flexible polymer chain is higher than that of the rigid cyclic monomer molecule and this entropy gain may overcome, at least for some systems, the loss of entropy due to the decrease of translational entropy. Elemental sulfur (8 – membered ring) is the best known example of this class of monomers. Some cyclic siloxanes and 6 – membered cyclic phosphates with large exocyclic groups also belong in this class.

$$\Delta H_p^o < 0; \quad \Delta S_p^o > 0$$

Polymerization would be possible over the entire temperature range at a standard concentration of $\mathbf{[M]_o = 1}$. At low monomer concentrations $\mathbf{\Delta G_p}$ may be positive.

$$\Delta H_p^o > 0; \quad \Delta S_p^o < 0$$

Polymerization is not possible. For small negetive $\mathbf{\Delta S_p}$ at high $\mathbf{[M]_o}$, may, however, $\mathbf{\Delta G_p^o}$ become negetive, making polymerization possible.

4.5.9 Molecular Weight

The molecular weight of the polymer is strongly dependent on the conditions of polymerization such as temperature, pressure, purity of monomer, etc. It has a very important bearing on the polymer crystallinity, kinetics of crystallization, morphology, rheology, physical properties and thermodynamic properties. Polymer almost contains invariably molecules of different chain lengths. The molecular weight control of polymers refers many of their properties such as physical properties depend on the chain length [55]. Knowledge of molecular weight and of the molecular weight distribution of a polymeric material is indispensable for scientific studies as well as for many technical applications of polymers. The methods developed for the determination of molecular weights subdivided into absolute and relative methods.

Polymer with higher molecular weight is almost independent of molecular weight. A slight increase in density occurs with increase in molecular weight. The highest percentage fraction of molecular weight of molecule will contribute the most toward the bulk density. High molecular weight polymer has high viscosity and poor processing ability. Molecular weight control and it distribution is often to obtain and improve desired physical properties in a polymer product.

4.5.9.1 Degree of polymerization

Degree of polymerization ($\overline{\mathbf{DP}}$), provides the size of polymer. It is a total number of structural units including end groups, and relates to both chain length and molecular weight. Polymer chains within a given polymer are always of varying lengths, it needs to use average value, such as number molecular weight, weight average molecular weight, etc.

The degree of polymerization and the molecular weight are some of the most important characteristics of a macromolecular substance. They indicate how many monomer units are linked to form the polymer chain and what their molecular weight is. In the case of homopolymers, the molecular weight of macromolecule is given by:

Molecular weight = degree of polymerization X molecular weight of the constitutional repeating unit

4.5.9.2 Number average molecular weight

The number-average molecular weight is the average molecular weight of a structural unit, times the average number of structural unit per chain with addition, the molecular weight of the end groups. The ratio of the converted monomer over the initiator provides the number average molecular weight ($\overline{\boldsymbol{Mn}}$) at any given conversion.

$$Mn\ (\mathrm{g/mol}) = \frac{\Delta \mathrm{M(g)}}{\mathrm{I(mol)}}$$

Where I = Initiator and M = Monomer.

Number average molecular weight ($\overline{\boldsymbol{Mn}}$) is the ratio between the total weight of all molecules (W) and the total number of molecules present.

$$\bar{M}n = W/\sum Nx = \sum Nx\, Mx/\sum Nx = \sum \bar{N}x\, Mx$$

Where

$$W = \text{Total weight of all molecules}$$

$$Nx = \text{Number of molecules of size } Mx$$

$$\bar{N}x = \text{Mole fraction of size } Mx$$

The degree of polymerization ($\overline{\mathbf{DP}}$), and the number average molecular weight, are important characteristics of polymer. They indicate polymer formation from the monomers link and their molecular weight. Number average molecular weight of polymer

$$\bar{M}n = \overline{DP}\,(Mo)$$

Number molecular weight of the polymer is the product of degree of polymerization and monomer after conversion is

$$\bar{M}n = \overline{DP}\,(Mo)$$

The weight and number average molecular weight in terms of P as conversion is

$$\bar{M}n = \frac{Mo}{1-P}$$

$$\bar{M}w = \frac{Mo\,(1+P)}{1-P}$$

4.5.9.3 Weight-average molecular weight

Weight fraction Wx of Mx molecules is the ratio between the weight concentration of Cx molecules and total weight concentration of all molecules C.

$$Wx = Cx/C$$

Weight concentration of Cx molecules is

$$Cx = Nx\, Mx$$

Where Nx is the number of molecules size Mx

$$C = \Sigma Cx = \Sigma Nx\, Mx$$

The weight average molecular weight is as follows

$$\overline{M}w = \Sigma Wx\, Mx = \Sigma Cx\, Mx / \Sigma Cx$$

Weight-average molecular weight is the average by weighting the molar mass of the molecules by the mass of each one present in the polymer.

$$\bar{M}_w = \frac{\sum_i N_i M_i^2}{\sum_i N_i M_i} = \frac{1}{m}\sum_i m_i M_i$$

where M_w = Weight-average molar mass

m_i = The total mass of molecules of molar mass M_i

m = The total mass of polymer

4.5.9.4 Viscosity average molecular weight

In the viscosity average molecular weight Mv, the weight average molecular weights and viscosity average molecular weights are equal when a is unity. In the equation given below, a is a constant which is greater for the larger sized polymer molecules than for smaller ones.

$$\overline{M}v = \left[\Sigma\, Mx^{a}\, Wx\right]^{1/a} = \left[\Sigma Nx\, Mx^{a+1} / Nx\, Mx\right]^{1/a}$$

The viscosity of a polymer in solution directly related to its molecular weight. The measurement of solution viscosity is relatively simple and does not require complex equipment. The results are appreciable between successive measurements. The solution viscosity measured from the intrinsic viscosity (η) by applying the Mark-Houwink equation given as

$$Mv = K\,(\eta)\,\alpha$$

Where K and α are constants depend on the type of polymer and nature of solvent [56].

4.5.9.5 Molecular weight distribution (MWD)

Any given polymer consists of several homologs differing only in the chain length *n*. There are smaller effects on crystallinity associated with average molecular weight and with molecular weight distribution. The effect of molecular weight distribution on crystallinity is slight based on the material of very low molecular weight. The control of molecular weight distribution used to obtain and improve desired physical properties in polymer product. Molecular weight or mass determines mechanical properties, viscosity and rheology.

MWD is one of the characteristic features of polymers. It is not critically a deficiency when polymers used as raw materials for products development. In case of uniform raw materials, the fundamental structures and properties of the polymers and polymerization mechanism can be advantage. Uniform polymers are very important for research in polymer chemistry. MWD is of significant importance on structure-property relationships in the solid state and in solution. This the substantial reason to concentrate on effort on preparation and chemistry of uniform polymers.

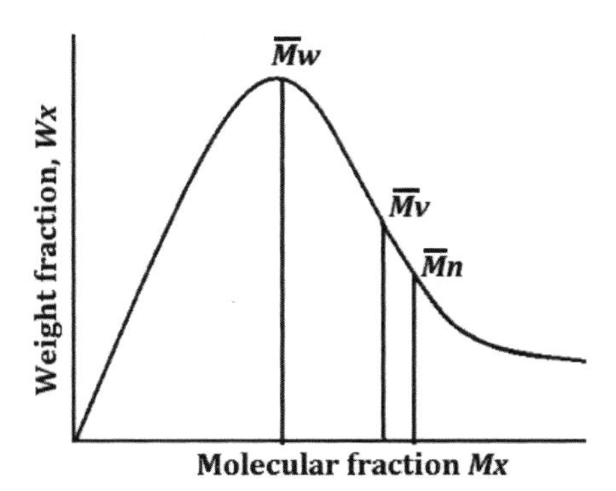

For a polydispersed polymer, Mw > Mv > Mn, with the differences between the various average molecular weights, increasing as the molecular weight distribution broadens. Description of material and polymers use molecular weight. These molecular weights characterized by various average molecular weights. Number and weight average molecular weights are the most frequently used. Any polymer the weight average molecular weight is greater than number average molecular weight $(\overline{M}_w > \overline{M}_n)$. Narrow molar mass distribution for polymer is often a higher degree in the solid and therefore higher density and melting point. Figure weight fraction Wx versus molecular fraction Mx indicates the in terms of weight, volume and number average molecular weight distribution. Polycondensates inherently exhibit a much more narrow distribution of MW compared to chain growth polymers.

The molecular weight distribution in step growth polymerization is the sum of the distribution of linear species and that of cyclic species in the polymerization of polyesters and polyamides. However, the linear polymers retain most probable distribution. Cyclic oligomers have different distribution [57–62].

4.5.9.6 Polydispersity and their index

Polymers are dispersed in nature. Any polymer material almost invariably contains molecules of different lengths due to the nature of polymerization. The difference between monodispersed and polydispersed is illustrated. Polymer's polydispersity is due to the chain formation during the polymerization. There are formations of various chain lengths present in the polymer [63].

The type of polymerization reaction controls the regularity of the polymer. Polymers are mixtures of molecules of different molecular weight in their pure form. Polydispersity lies in the statistical variation present in the polymerization process. It is true for common polymerization reaction such as chain growth and step growth polymerizations. Polydispersity index provides polydispersity of the polymer.

In polycondensation reaction, the regularity of chain structure is much easier to control due to the equal participation of monomer with functional groups. In polyaddition reaction, the nature of propagation reaction allows variation in the addition of monomer to the growing chain. Even the termination reaction results in greater polydispersity of molecular weight due to variability in the reaction also.

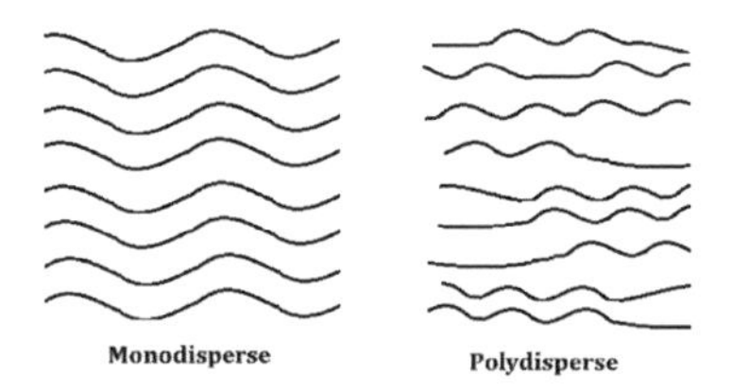

For monodispersity $\overline{M_w} = \overline{M_n}$

Polydispersity index $= \overline{M_w}/\overline{M_n}$

Degree of polymerization $\overline{DP} = 1/(1-P)$

At a given conversion, the polydispersity index is

$$\text{Polydispersity index} = \overline{\mathbf{M}}_{\mathbf{w}}/\overline{\mathbf{M}}_{\mathbf{n}}$$

The polymers produced have a polydispersity described by the Poisson distribution. Therefore, the polydispersity index is [64].

$$\overline{\mathbf{M}}\mathbf{w}/\overline{\mathbf{M}}\mathbf{n} = 1 + P$$

$$\overline{\mathbf{M}}\mathbf{w}/\overline{\mathbf{M}}\mathbf{n} \geq 1$$

Polydispersity = 1.02 or 1.05 is to be considered as monodisperse.

The most frequently utilized averages are the number and weight average molecular weights. It can be shown that for any polymer $\mathbf{M_w} > \mathbf{M_n}$, except for a monodispersity of a polymer $\mathbf{M_w} = \mathbf{M_n}$. $\mathbf{M_w}/\mathbf{M_n}$ ratio is a good indicator of the polydispersity of a polymer [65].

4.5.10 Glass Transition Temperature (T_g)

Glass transition temperature often takes place over a broad range of temperatures. The glass transition temperature is manifested in a polymeric material as an embrittlement when moving across the range from high to low temperatures. Polymer with low molecular weight behaves as fluid above its T_g. High molecular weight polymer behaves as rubber above its T_g.

The glass transition temperature intervenes at amorphous regions, largely through controlling the main chain stiffness. Large changes occur in properties such as chemical reactivity, mechanical and dielectric relaxation, viscous flow, load-bearing capacity, hardness, tack, heat capacity, refractive index, thermal expansion, creep and diffusion during it traverses in glass transition range of temperature.

4.6 Material Considerations

Appropriate application for a given polymer is to determine numerous considerations. For a polymer, it should be commercially viable along with stringent environmental regulations, high performance standards and more over economical. The properties of polymers depend on their basic chemical and structural properties as mentioned below:

- Molecular weight – the molecular weight and their distribution strongly affect the properties of polymers and their applications.

- Hydrophobic or hydrophilic – the type and content significantly impacts polymer inter- and intra-molecular associations.
- Sequence distribution of monomers – in case of block, alternating, or random distribution affects interaction of monomer groups within the polymer.
- Degree of branching – a branched polymer has different properties such as low tendency to crystallize, etc. Branching offer the synthetic ability to tailor chemically the branches in a stepwise fashion.
- Degree of cross-linking – cross-linked polymer has chemical linkages between chains. In the presence of solvent, it usually swells, but does not dissolve.
- Ionic character – depends on the number of charged groups, charge type and charge distribution.
- Chemical modification – chemically modified to adopt their properties to the needs of a particular application.

4.7 Summary

- Polymer chemistry has become a mature science. It has all the advantages and handicaps of maturity.
- Polymers are substances containing a large number of structural units joined by the same type of linkage. The substances form into a chain like structure.
- Polymer chains tend to be flexible and easily entangled or folded.
- Degree of crystallinity is the amount of ordering in a polymer. Many of the properties based on the degree of crystallinity.
- Variations in properties cause changes in average molecular weight. The changes in molecular weight are due to differences in the polymer structure.
- Instead of a single molecular weight, polymer synthesis is as a result of dealing with a molecular weight distribution (MWD).
- The chain may have different end group chemistries due to different reaction processes, thereby creating a functionality type distribution.
- In random copolymers, the polymer chains show in addition a chemical composition distribution.
- In block copolymers, addition sequence and block-length distribution are present.
- Polymers and copolymers show an architecture distribution (linear, cyclic, branched, etc.).

- Stretching or extruding a polymer can increase crystallinity
- Polymers growth to be larger with applications that far exceeds towards any other class of materials available to human kind.

References

[1] G. Gruenwald. *Plastics, how structure determines properties, Hanser*, Munich (1992).
[2] E. H. Andrews. *Fracture in polymers, Oliver and Boyd,* London (1968).
[3] I. Masaru, Narisawa, I. and H. Ogawa. *Poly. J.* **8** 391–400 (1976).
[4] T. Alfrey, J. J. Bohrer, and H. Mark. *Copolymerization, Interscience*, New York (1952).
[5] P. Cigana, B. D. Favis, and R. Jerome. *J. Polym. Sci. Polym. Phys.* **34**, 1691–1700 (1996).
[6] S. Lyu, T. D. Jones, F. S. and Bates, C. W. Macosko, *Macromolecules.* **35**, 7845–7855 (2002).
[7] N. C. Tan, Tai, S. K. and Briber, R. M. *Polymer*, **37**, 3509–3519 (1996).
[8] S. T. Milner, and H. Xi, *J. Rheol.* **40**, 663–687 (1996).
[9] J. C. Lepers, and B. D. Favis, *AIChE J.* **45**, 887–895 (1999).
[10] M. Matos. B. D. Favis, and P. Lomellini. *Polymer* **36**, 3899–3907 (1995).
[11] P. Cigana, B. D. Favis, and R. Jerome. *J. Polym. Sci., Polym. Phys.* **34**, 1691–1700 (1996).
[12] J. Noolandi, and K. M. Hong. *Macromolecules* **15**, 482–492 (1982).
[13] P. Cigana, B. D. Favis, R. Jerome. *J. Polym. Sci., Polym. Phys.* **34**, 1691–1700 (1996).
[14] M. Matos, B. D. Favis, and P. Lomellini. *Polymer* **36**, 3899–3907 (1995).
[15] K. R. Shull, E. J. Kramer, G. Hadziioannou, and W. Tang, *Macromolecules* **23**, 4780–4787 (1990).
[16] M. J. Arlen, and M. D. Dadmun, *Polymer* **44**, 6883–6889 (2003).
[17] R. E. Falk, L. Makowka, N. Nossal, J. A. Falk, J. E. Fields, and S. S. Asculai, *Br. J. Sugery*, **66**, 861 (1979).
[18] *Br. Pat.* Anm GB 2040951A.
[19] S. Paraskeva, N. Hadjichristidis, *J. Polym. Sci., Part A: Polym. Chem.* **38**, 931–935 (2000).
[20] N. Hadjichristidis, M. Pitsikalis, S. Pispas, H. Iatrou. *Chem. Rev.* **101**, 3747–3792 (2001).
[21] D. Uhrig, and J. W. Mays. *Macromolecules* **35**, 7182–7190 (2002).
[22] N. Haraguchi, and A. Hirao. *Macromolecules* **36**, 9364–9372 (2003).

[23] P. F. Rempp, and P. J. Lutz. In *Comprehensive Polymer Science*, Eastmond, G. C., Ledwith, A., Rasso, S., Sigwalt, P., Eds, Pergamon Press: New York, **6**, 403 (1989).
[24] H. L. Hsieh, and R. P. Quirk. *Anionic Polymerization: Principles and Applications*, Marcel Dekker, New York, 369–392 (1996).
[25] I. M. Ward *Mechanical properties of solid polymers*, 2nd edn. John Wiley, New York (1983).
[26] G. Gruenwald. *Plastics, how structure determines properties. Hanser*, Munich (1992).
[27] A. V. Tobolsky and R. D. Andrews. *J. Chern. Phys.* **13**, 3 (1945).
[28] M. M. Denn. Extrusion Instabilities and Wall Slip. *Annu. Rev. Fluid Mech.* **33**, 265 (2000).
[29] T. Alfrey, J. J. Bohrer, and H. Mark. *Copolymerization, Interscience*, New York (1952).
[30] J. D. Ferry. *Viscoelastic properties of polymers*, 3rd edn. John Wiley, New York (1990).
[31] K. Matsumoto, J. F. Fellers, and J. L. White, *J. Appl. Polym. Sci.* **26**, 85 (1981).
[32] J. E. Flood, J. L. White, and J. F. Fellers, *J. Appl. Polym. Sci.* **21**, 2965 (1982).
[33] H. Bodaghi and T. Kitao. *Polym. Eng. Sci.* **24**, 242 (1984).
[34] H. Pasch Hyphenated techniques in liquid chromatography of polymers. In *Advances in Polymers Science, New Developments of Polymer Analytics*. I. Berlin: Springer, 1–66 (2000).
[35] H. J. A. Philipsen. Determination of chemical composition distributions in synthetic polymers. *J. Chromatogr. A* **1037**, 329–50 (2004).
[36] F. Rybnikar. *Macromol Sci-Phys* **B19**, 1 (1981).
[37] F. Rybnikar. *J Appl Polym. Sci.* **27**, 1479 (1982).
[38] H. Schnell. *Chemistry and physics of polycarbonates. John Wiley Interscience*, New York (1964).
[39] J. H. Magill. In Schultz JM (ed) *Treatise on materials science and technology*, **10**. Academic Press, New York, 3 (1977).
[40] A. S. Vaughan, and D. C. Bassett. *Polymer* **29**, 1397 (1988).
[41] E. Tromsdorff, H. Kolile, and P. Lagully, *Makromol. Chem.* **1**, 169 (1948).
[42] H. B. Lee, and D. T. Turner. *Macromolecules* **10**, 226 (1977).
[43] N. Friss, and L. Nykagen. *J. Appl. Polym. Sci.* **17**, 2311 (1973).
[44] D. T. Turner. *Macromolecules* **10**, 221 (1977).

[45] C. E. Carraher, Jr. Polymer Chemistry, sixth ed. *Revised and Expanded. Marcel Dekker Inc.*, New York (2003).

[46] R. W. Jones, and R. H. M. Simon. Synthetic Plastics. In: Kent, J.A. (Ed.), *Riegel's Handbook of Industrial Chemistry,* eighth ed. Van Nostrand Reinhold, New York (1983).

[47] E. Lokensgard. *Industrial Plastics: Theory and Applications. Delmar Cengage Learning, Clifton Park*, New York (2010).

[48] E. Turska, and M. Oblój-Muzaj. *Acta Polymerica* **32,** 295–99 (1981).

[49] C. De Rosa and F. Auriemma. *J. Am. Chem. Soc.* **128**, 11024–11025 (2006).

[50] Dosch, H. *Critical Phenomena at Surfaces & Interfaces*, 1st ed.; Springer-Verlag, Berlin, (1992).

[51] K. S. Gautam, and A. Dhinojwala. *Phys. ReV. Lett.* **88**, 1–4 (2002).

[52] K. S. Gautam, S. Kumar, D. Wermeille, D. Robinson, and A. Dhinojwala. *Phys. Rev. Lett.* **90**, 1–4 (2003).

[53] S. Prasad, L. Hanne, and A. Dhinojwala. *Macromolecules* **38**, 2541–2543 (2005).

[54] B. M. Ocko, X. Z. Wu, E. B. Sirota, S. K. Sinha, O. Gang, M. Deutsch. *Phys. Rev. E* **55**, 3164–3182 (1997).

[55] O. Gang, X. Z. Wu, B. M. Ocko, E. B. Sirota, and M. Deutsch. *Phys. Rev. E* **58**, 6086–6100 (1998).

[56] A. Valdebenito, and M. V. Encinas. *Polymer* **46**, 10658–10662 (2005).

[57] H. W. Melville. *Proc. Roy. Soc (London)* **A237**, 149 (1956).

[58] H. Jacobson, W. H. Stockmayer. *J. Chem. Phys.* **18**, 1600 (1950).

[59] H. Jacobson, C. O. Beckmann, and W. H. Stockmayer. *J. Chem. Phys.* **18**, 1607 (1950).

[60] P. J. Flory, J. A. and Semlyen. *J. Am. Chem. Soc.* **88**, 3209 (1966).

[61] P. J. Flory, U. Suter, M. and Mutter, *J. Am. Chem. Soc.* **98**, 5733 (1976).

[62] M. Mutter, U. W. Suter, P. J. and Flory. *J. Am. Chem. Soc.* **98**, 5745 (1976).

[63] U. W. Suter, M. Mutter, and P. J. Flory. *J. Am. Chem. Soc.* **98**, 5740 (1976).

[64] J. A. Semlyen. *Adv. Polym. Sci.* **21**, 41–75 (1976).

[65] J. E. Puskas, and G. Kaszas. *Prog. Polym. Sci.* **25**, 403–452 (2000).

[66] I. G. Voight-Martin, E. W. Fischer, and L. Mandelkern *J. Polym. Sci. Polym. Phys.* **18**, 2347 (1980).

5

Step Growth Polymerization

Step growth polymerizations are of industrial importance. Monomers having two different groups of atoms such as ester or amide can join to form condensation polymer with the release of small molecules such as water, ammonia, etc. Polyesters and polyamides are important commercial polymers. Step growth polymerization reactions are condensation polymerization of difunctional monomers. The stoichiometric associated with the formation or not formation of small molecules during the reaction. Polyester and polyamides are industrially important polymers by step growth polymerization. Phenol and aldehyde reaction is a step growth polymerization. Chain polymerization is an addition polymerization of olefinic monomers. Styrene reaction is a chain polymerization [1–4].

5.1 Functional Groups

In step growth polymerization, functional groups are involved in the reactions. The monomer starts can be either bifunctional or multifunctional. In step growth polymerization, the reaction through functional groups occurs. The linear structure forms with bifunctional monomer of either branched or network structure with multifunctional monomer. Two monomers of same or different functional groups or bifunctional monomers involve in the reaction. The monomers need not necessarily add sequentially. In the polymerization, even small polymer chains may couple into larger chains. Functional groups such as –COOH, –NH2, –OH, etc., either one or more are located in the molecules. The growth center can occur by the reaction between two molecules whether polymeric or monomeric.

The functional group polymerization is termed as step growth reaction. It is based on the reaction of two functional groups such as an acid and an alcohol to form an ester or the reaction of an amine and an acid to produce amide, etc. This reaction evolves low molecular mass species, such as water, methanol, hydrogen chloride etc. Therefore, this reaction is called polycondensation reaction. The polymerizations of phthalic anhydride with diols, diisocyanates with diols, etc., are of step growth polymerization even though they are not condensation polymerization.

5.2 Reactions

Step growth polymerization is an important method of polymerization involved in synthesis and manufacture of polyamides, polyesters, polyimides and π-conjugated polymers.

- With two different monomers
  Different mutually reactive groups occur not on the same but on different molecules mentioned in Equation R1.

$$nNH_2(CH_2)_6NH_2 + nHOOC(CH_2)_4COOH \longrightarrow H{-}\left[NH(CH_2)_6NHOC(CH_2)_4CO\right]_n{-}OH + (2n-1)H_2O \quad \text{R1}$$

Hexamethylene diamine; Adipic acid; Nylon

- One monomer with two different functional groups
  One molecule carries on either end or the two dissimilar reactive groups (e.g. ε aminocaprioic acid) as shown in Equation R2.

$$n\,NH_2(CH_2)_5COOH \longrightarrow H{-}\left[NH(CH_2)_5CO\right]_n{-}OH + (n-1)\,H_2O \quad \text{R2}$$

ε aminocaproic acid

- Retains their functionality as end groups at the completion of polymerization.
  By a reagent abstracts combines with the end groups from the reacting and polymerizing molecules. Decamethylene dibromide in the presence of sodium enables the residues to join up with the elimination of sodium bromide as mentioned in Equation R3.

$$Br(CH_2)_{10}\boxed{Br + Na} + \boxed{Na + Br}(CH_2)_{10}\boxed{Br + Na} + \boxed{Na + Br}(CH_2)_{10}Br \quad \text{R3}$$

$$\longrightarrow Br(CH_2)_{10}(CH_2)_{10}(CH_2)_{10}Br + NaBr$$

5.3 Polymerization

Based on the functional groups present in the monomers, functionality of the monomers controls the structure as linear, branched, or cross-linked macromolecules. The commercially important polymers are created by the reaction leading to the formation is called polymerization. Among all well-known step polymerization, condensation polymerizations are far more common. This also applies to reactions that have industrial importance. Some of the well-known systems of monomers of Step polymerizations include condensation and addition. These reactions are with monomers with two functional groups. The difference between condensation and addition lies in the fact that in the reaction, growth of the polymer chain accompanies by the formation of low molecular weight molecules such as ammonia, hydrogen chloride, water as condensation by-products. The removal of such small molecules shifts the reaction equilibria and is necessary for the progress of the reaction. Even though polyesters and polyurethanes are not condensation polymerizations. Recently these polymerizations have been classified into random and sequential types [5–6]. The side reaction products even though in small concentrations lead to important changes in the physical properties. Therefore, these side products affect the end use of the polymer [7].

5.3.1 Step Growth Polymers

Step growth polymers form with elimination of a small molecule. The small molecule differs depending upon monomers used to form the polymer. Step growth polymerization undergoes reactions to form higher molecular weight products. This polymerization include cationic and anionic aggregation reactions. The reactions occur in either aqueous or non-aqueous or at high temperatures in absence of solvents. This polymerization also occurs when hydrated, ammoniated or solvated products which undergo dehydration or desolvation at high temperatures. Elimination of low molecular weight species brings almost condensation reactions on heating. Also partially, solvated intermediates undergo polymerization by elimination of low molecular species. Interactions of functional groups in molecules can conceivably modified by replacing one of the reactants with other functional compound of either organic or inorganic in nature.

Linear chain obtained by the step growth reaction by intermolecular condensation or addition of the reactive groups in bifunctional monomers. The reaction occurs with or without elimination of small molecules such as water, ammonia, etc.

- Monomers join up with expulsion of small molecules
- Not all the original atoms are present in the polymer

5.4 Polyaddition Reactions

5.4.1 Transesterification

Transesterification and the product formed from dimethylterephalate and ethylene glycol mentioned in Equation R. Dimethylterephate dissolves in ethylene glycol and react in the presence of suitable catalyst. The methanol condenses evaporate along with ethylene glycol and is further recycled.

$$CH_3OC(=O)-C_6H_4-C(=O)OCH_3 + 2HOCH_2CH_2OH$$

Dimethyl terephthalate　　Ethylene glycol

$$\longrightarrow HOCH_2CH_2OC(=O)-C_6H_4-C(=O)OCH_2CH_2OH + 2CH_3OH \dashrightarrow R$$

Bis-hydroxyethyl terephthalate　　Methanol

$$\sim C_6H_4-C(=O)OCH_3 + \sim C_6H_4-COCH_2CH_2OH$$

$$\longrightarrow \sim C_6H_4-C(=O)-OCH_2CH_2OC(=O)-C_6H_4\sim + CH_3OH$$

5.4.2 Ester-Interchange

Ester-interchange reaction is used to prepare poly(ethylene terephthalate) and related esters. The order of reactivity is as follows: PBT considered as a superior polymer because of its outstanding properties. It produced by ester interchange between dimethyl terephthalate and butanediol, while PET is derived from ethylene glycol as the diol component.

Terephthalic acid > isophthalic acid > phthalic acid > benzoic acid ester
Whereas sebacic acid is as reactive as terephalic acid.

$$\sim C_6H_4-C(=O)OCH_3 + HOCH_2CH_2OH \longrightarrow \sim C_6H_4-C(=O)OCH_2CH_2OH + CH_3OH$$

5.4.3 Polyesterification

The polyesters prepared by the polymerization of glycols with either anhydrides or acids. Unsaturated polyesters are prepared by using unsaturated anhydride (for example, maleic anhydride) or a diacid (fumaric acid). The polyesterification in presence of ethylene glycol occurs as Terephthalic acid undergoes reaction with ethylene glycol form poly(ethylene terephthalate) with elimination of water.

$$n\ HO-\overset{O}{\overset{\|}{C}}-C_6H_4-\overset{O}{\overset{\|}{C}}-OH + n\ HO-CH_2CH_2-OH \longrightarrow$$

terephthalic acid ethylene glycol

$$\left[-O-\overset{O}{\overset{\|}{C}}-C_6H_4-\overset{O}{\overset{\|}{C}}-O-CH_2CH_2-\right]_n + 2n\ H_2O$$

poly(ethylene terephthalate)

The polyesterification can take place in the presence of an external catalyst. In the absence of this, the diacid monomer acts as its own catalyst for the reaction. The polyesterification reaction can take place in the presence or in the absence of switches of chains, too.

5.5 Polycondensation

Industrial polycondensation or polyaddition reactions, [8] ensure that the exact equimolar ratio of the two reactants is necessary to achieve high molecular weights where it finally reaches.

$$2 \sim C_6H_4-COCH_2CH_2OH \longrightarrow$$

$$\sim C_6H_4-\overset{O}{\overset{\|}{C}}-OCH_2CH_2O\overset{O}{\overset{\|}{C}}-C_6H_4\sim + HOCH_2CH_2OH$$

Polycondensation has simplicity, rich synthetic possibilities, and low tendency to the undesired chain-termination side reactions with solvents and tertiary amines, necessarily used as acid acceptors. The reaction proceeds in mild conditions, and the reactivity of eletrophilic monomers (and, consequently, polycondensation processes) regulated in a wide range by using different pendent groups, the number and nature of which are virtually unlimited. This opens unique possibilities to polycondensation. In the case of Polycondensation, the formation of polymer undergoes hydrolysis of reactive acid chlorides or isocyanates results in a loss of stoichiometry in the reaction process and forms urea instead of polyurethane.

Major factors affecting low-temperature polycondensations are:

- Reaction rate
- Purity of intermediates and solvents
- Equivalence of reactants
- Solubility of intermediates and polymer
- Mixing
- Side reactions
- Temperature

5.6 Characteristics

Step polymerizations have only one type of reaction. It is the reaction of chain growth. Both chain and step polymerizations have no difference with initiation, propagation, and termination reactions. With respect to chain polymerization, step polymerization proceeds much more slowly. The polymer with a high degree of polymerizations achieve only at very high degree

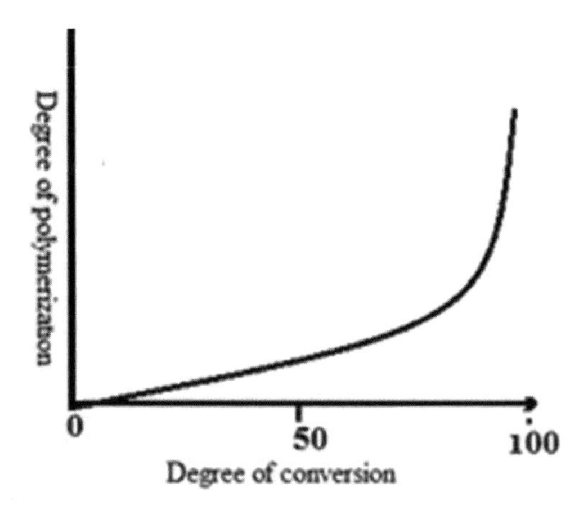

Figure 5.1 Relation between degree of polymerization vs conversion – step growth polymerization.

of conversion. The slow increase of the degree of polymerization follows with monomers transform into dimers, then into trimers, tetramers and so on. Finally, the polymer obtains in the reaction mixture.

Step growth polymerization is an important method of polymerization that yields not only engineering polymers such as polyamides, polyesters, and polyimides but also π-conjugated polymers. The molecular weight of those polymers is generally difficult to control, and the polydispersity index theoretically approaches two at high conversion, which is unlike the behavior of living polymerization. An uncontrolled molecular weight and broad molecular weight distribution do not stem inherently from the reaction type of condensation polymerization, i.e. condensation steps with elimination of a small molecule [9].

In linear polymerization X–X and Y–Y, one must adjust the stoichiometric balance of the bifunctional monomers. The number average degree of polymerization $\bar{X}n$ with stoichiometric unbalanced r at complete conversion is

$$\bar{X}_n = \frac{(1+r)}{(1-r)}$$

The relation between degree of polymerization versus conversion of monomers into polymers is as shown in Figure 5.1. The polydispersity index (PDI) of a heterogeneous step growth polymerization, where the solubility of A – A is limited, as a function of unreacted B–B functional groups [10].

5.7 Elementary Step Growth Polymerization

Bifunctional monomers of the XRX and YR'Y react forms linear step growth polymerization. The reaction between terephthalic acid and ethylene glycol in polyester formation is an example of the monomers of XRX and YR'Y type reaction [11].

Linear polymer chains in step growth polymerization, the starting monomer must be bifunctional. The starting monomer X and Y can be on the same molecule as in X–R–Y, where R is an alkyl or aryl group. Aminocaproic acid [HOOC – $(CH_2)_5$ – NH_2) consisting of functional groups –COOH and –NH2 that can react to give a –CONH– linkage in polymer and hexamethylene adipamide commonly known as Nylon 6,6 salt the concentration of functional groups X and Y are equal. In this P_m and P_n denote polymer having m and n monomeric units, respectively. The reaction proceeds mechanistically in several steps [12–14].

$$X_m + Y_n \underset{K'_{p,m+n}}{\overset{K_{p,mn}}{\rightleftharpoons}} P_{m+n} + C$$

Where m = 1, 2, ...
n = 1, 2, ... and m+n = 2, 3, ...

X and Y are monomers with difunctional groups of numbers m and n respectively. P_{m+n} is a polymer molecule of length (m+n) formed as a result of the reaction between the functional groups and C is the condensation product P_m and P_n. $K_{p,mn}$ is expected to be complex function of the chain length m and n of the molecules involved. The reverse reaction involves a large and a small molecule, and the rate constant $K'_{p,m+n}$, is a function of the chain length of P_{m+n} only.

According to LeChatlier's principle removal of by-product small molecules such as water, ammonia, etc., the reaction drive towards forward.

$$X_m + Y_n \underset{K_r}{\overset{K_f}{\rightleftharpoons}} P_{m+n} + C$$

Where X_m and Y_n are monomers, P_{m+n} and C are polymers and by product and b are monomers, c and d are polymer and by-product, and k_f and k_r are kinetics of forward and reverse reactions.

In the reaction mass, molecules of X and Y diffuse to each other and interact to form a chemical bond. Segmental diffusion are dependent on the chain lengths of the two polymer molecules in the reaction. Flory verified the kinetics of catalyzed and non-catalyzed polyesterification of –COOH and –OH group [15].

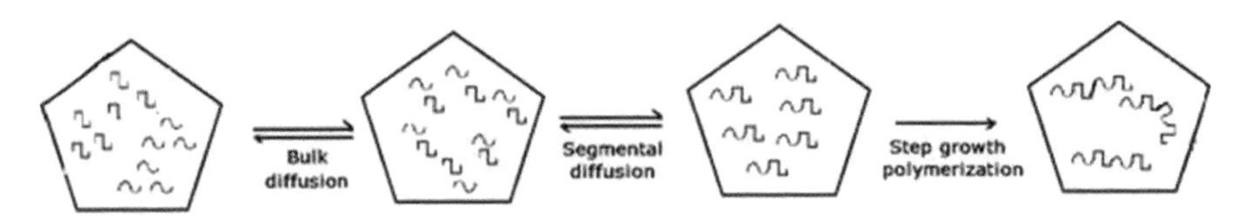

5.8 Kinetics

Step growth polymerization of difunctional monomers –X–R–Y where X and Y are mutually reactive functional groups. The polymerization occurs with the elimination of small molecule as by-products of XY, a linear polymerization with general formula of X–(R)$_n$–Y is formed [16–18].

$$X\text{-}R\text{-}Y \xrightarrow[\text{single-bond reaction}]{\text{step polymn}} X{+}R{+}_nY \quad \text{Condensation polymer}$$

5.8.1 Step Growth Polymerization

Where X – R – Y is a difunctional monomer and X and Y are reactive functional groups. Step growth polymerization is a single bond reaction.

In A-A plus B-B step-growth polymerization, a linear chain obtained by the stepwise intermolecular condensation or addition of the reactive groups in bifunctional monomers. The molecular weight distribution (MWD) of step-growth systems are analyzed by Flory with the assumption that all functional groups are considered as being equally reactive [19]. Three types of molecules with respect to the end-group functionality, which form in the polymeric mixture.

$$A\text{–}A \quad + \quad B\text{–}B \longrightarrow \text{+}A\text{–}AB\text{–}B\text{+}_n$$

If n, the total number of reactant molecules combined in the polymer molecule, is an integer,

$$\left(\frac{n}{2}\right)A\text{–}A + \left(\frac{n}{2}\right)B\text{–}B \longrightarrow A\text{–}A\left(B\text{–}BA\text{–}A\right)_{(n-2)/2}B\text{–}B$$

If n is odd, either

$$\left(\frac{n+1}{2}\right)A\text{–}A + \left(\frac{n-1}{2}\right)B\text{–}B$$
$$\longrightarrow A\text{–}A\left(B\text{–}BA\text{–}A\right)_{(n-3)/2}B\text{–}BA\text{–}A$$

or

$$\left(\frac{n+1}{2}\right)A\text{–}A + \left(\frac{n-1}{2}\right)B\text{–}B$$
$$\longrightarrow B\text{–}B\left(A\text{–}AB\text{–}BA\text{–}A\right)_{(n-3)/2}B\text{–}B$$

There is a slow increase in the average molecular weight of the polymer due to the growth process is random in nature, and presence of oligomers with different chain length at any instant of time. Molecular weight of the polymer from step-growth polymerization are relatively small. Small amounts of unreacted monomer presents even after beginning of the reaction.

In step growth polymerization the kinetics of ester formation is very difficult to determine from experimental measurement of rate of polymerization alone [20–21].

5.8.2 Uncatalyzed Polymerization

In uncatalyzed polymerization, the acid end groups are found to act as a catalyst. As a result, the overall rate of polymerization as second order in [COOH] and first order in [OH].

The overall rate of polymerization is as below

$$R_p = -\frac{d[\mathrm{COOH}]}{dt} = k_1[\mathrm{COOH}]^2[\mathrm{OH}]$$

In case of equal moles of acid and hydroxyl group in the feed, the above expression reduces to a third order rate expression,

$$-\frac{d[\mathrm{COOH}]}{dt} = k_1[\mathrm{COOH}]^3$$

Integration of the equation, the functional group conversion P_A derived as

$$\frac{1}{(1-P_A)^2} = 2[\mathrm{COOH}]_0^2 K_1 t + \text{Constant}$$

Where $[\mathbf{COOK}]_0$ **= the initial acid group concentration** in the reaction mass.

Being the reaction is slower due to time dependence. It is necessary to drive reaction with the addition of acid.

- With high reaction temperature increases k and removal of by-product such as evaporation of H_2O.
- Without solvent such as mass or bulk polymerization case, $[M]_0$ is maximum and with no need to separate product. Also with high π, the viscosity η can be low along with improvement of heat and mass transfer.

- With solvent, the monomer are not miscible with each other. However, miscible with solvent and allow high temperature. Being with solvent, the medium can be dilute viscous media. Therefore with solvent improvement of processing with heat and mass transfer improvement.

5.8.3 Catalyzed Polymerization

Polymerization carried out in the presence of a suitable catalyst, the rate of polymerization is

$$R_p = -\frac{d[COOH]}{dt} = k_1' [COOH][OH]$$

The equation on integration, for an equal molar ratio of COOH and OH groups yields.

$$\ln \frac{1}{1-P_A} = [COOH]_0 k_1' + \text{Constant}$$

A slow reaction with –COOH group decreases and gets extremely slow at the end of the reaction.

For nearly an equilibrium step growth polymerization

- With higher temperature, k is to increase. In such case, k increases with temperature increases, therefore, it helps to remove by-product.

In bulk or mass polymerization without solvent

- In case of miscible reactants, which one uniform mixed phase, which gives highest possible higher concentration.
- Being without solvent, there is no need to have separation step to remove solvent.
- To have higher molecular weight, the viscosity η low enough to process. Therefore, the product directly processed into final form.

With solvent

- In case of solvent in the reaction, the two monomer present as reactants need to solubilize.
- Being with solvent, the reaction could allow approaching higher temperature without degrading the polymer.
- For viscous medium, the solvent carrier helps exceptionally with higher molecular weight.

5.8.4 Polyesterification Using Acid Catalyst

In polyesterification, acid catalyst is more reactive than –COOH group which involve in the reaction. The general kinetic expression for the polyesterification of adipic acid and ethylene glycol should be as below [22–24].

$$\sim\!\!\sim C(=O)\!-\!OH + HA \underset{k_2}{\overset{k_1}{\rightleftharpoons}} \sim\!\!\sim \overset{\oplus}{C}(OH)\!-\!OH\ A^{\ominus}$$

$$k_{1,2} = \frac{k_1}{k_2} = \frac{\left[\sim\!\!\sim \overset{\oplus}{C}(OH)\!-\!OH\ A^{\ominus}\right]}{[HA][COOH]}$$

5.8.4.1 Slow step rate-determining step

The hydrogen ions concentration in the medium of the reaction mass due to the ionization of the acid, only small quantities of acid dissociated. This ion concentration does not depend upon the total carboxylic acid in reaction with hydroxyl group added. However, the ion concentration depends on the hydroxyl group in reaction with carboxylic acid in the polyesterification reaction.

$$\sim\!\!\sim \overset{\oplus}{C}(OH)\!-\!OH\ A^{\ominus} + \sim\!\!\sim OH \underset{k_4}{\overset{k_3}{\rightleftharpoons}} \sim\!\!\sim C(OH)(OH)\!-\!\overset{\oplus}{O}H\!\sim\!\!\sim\ A^{\ominus}$$

Rate of disappearance of carboxylic monomer

$$R_p = K[COOH][OH][H^+] = -\frac{d[COOH]}{dt}$$

$$R_p = k_3\,[C^+(OH)_2]\,[OH]$$

$$= k_3\,k_{1,2}[HA][COOH][OH]$$

5.8.4.2 Catalyst regeneration

$$\sim\!\!C(:\ddot{O}H)(\ddot{O}H)(\overset{\oplus}{O}H\!\sim)\; A^{\ominus} \underset{}{\overset{K'}{\rightleftharpoons}} \sim\!\!C(OH)\!-\!O\!-\!\sim + H_2O + HA$$

From the equilibrium concentration, [HA] is constant due to regeneration of the catalyst.

$$R_p = -\frac{d[COOH]}{dt} = K'[COOH][OH]$$

where $K' = k_3 k_{1,2}[HA]$

In other word

$$[H^+] = K_\alpha [OH]$$

where K_α = **proportionality constant**

If the molar ratio is fed in the batch reactor is r, then

$$r = \frac{[OH]_o}{[COOH]_o}$$

The equation reduces to

$$-\frac{d[COOH]}{dt} = K_a[COOH](a + [COOH])^2$$

Where

$$-\frac{d[COOH]}{dt} = K_a[COOH](a + [COOH])^2$$

$$K_a = \kappa K_\alpha$$

$$a = -COOH \text{ groups}$$
$$= (r-1)[COOH]_o$$

The equation after integration becomes as below

$$\ln \frac{r - P_A}{1 - P_A} - \frac{r-1}{1-P_A} = a^2 K_a t + \left(\ln r - \frac{r-1}{r}\right)$$

where $P_A = -$**COOH groups conversion**

5.9 Biodegradability

Biodegradable nature of step growth polymers generally depends on

- Chain coupling – ester > ether > amide > urethane
- Molecular weight becomes lower and degradation is faster than higher molecular weight polymer
- Morphology prefers amorphous than crystalline
- Hydrophilicity is faster than hydrophobicity

5.10 Summary

- A reaction through functional groups occurs and it occurs only either at least with two functional group molecules or monomers with two functional groups. They present oligomers in different chain length at any instant of time.
- Monomers need not add sequentially, however, small polymer chain may couple into larger chains. Branched or cross linked polymers are obtained when more than two functional groups in the monomer molecules are present. With a bifunctional monomer, linear polymers formed.
- Monomers disappear in the initial stage. Monomer or polymer can react even at any molecular weight. Molecular weight of the polymers are relatively small. Small amounts of unreacted monomer presents even after beginning of the reaction.
- The growth center can occur by the reaction between two molecules whether polymeric or monomeric. Slow increase in the average molecular weight of the polymer due to the growth process is random in nature. All molecular species present throughout reaction.
- Slow molecular weight increases at initial stage, then increases rapidly at high conversion. Molecular weight increases with long reaction time. Yield hardly changes.

- Molecular weight control in linear polymerization is crucial in achieving the balance of mechanical properties and processibility. The two most common methods for molecular weight control are the adjustment of the concentration of the two monomers so that they are slightly non-stoichiometric and the addition of a small amount of a monofunctional monomer.
- Accurate transfer of the monomers into a reactor can be challenging, especially when appropriate solvents used to rinse the weigh containers [25].
- Polyaddition and polycondensation reactions usually lead to a large number of side products and only to oligomers in the aqueous phase, these new conditions resulted in high molecular-weight, defect-free polymers.

References

[1] C. E. Carraher. *Seymour/Carraher's Polymer Chemistry*. CRC Press, Boca Raton, FL (2008).

[2] W. A. Braunecker, K. and Matyjaszewski. *Prog. Polym. Sci.* **33**, 165 (2008).

[3] Y. Yoon, R. M. Ho. F. M. Li. M. E. Leland, J. Y. Park. S. Z. D. Cheng, V. Percec, and P. W. Chu. *Prog. Polym. Sci.* **22**, 765 (1997).

[4] P. C. Hiemenz, and T. P. Lodge. *Poly. Chem.* CRC Press, Boca Raton, FL, (2007).

[5] D. Fabbri, C. Trombini, I. and Vassura, J. *Chromatogr. Sci.* **36** 600 (1998).

[6] J. C. del Rio, A. Gutiérrez, F. J. Gonzales-Vila, and F. Martin, *J. Anal. Appl. Pyrol.* **49** 165 (1999).

[7] R. Darryl, F. D. Harvey, H. E. Carlton. Ash and R. Dwayne. *Senn. Macromol.* 30, 387–398 (1997).

[8] A. Kumar, R. Gupta. *Fundamentals of Polymer Engineering*. (2nd ed.) Marcel Dekker, New York, (2003).

[9] T. Yokozawa and A. Yokoyama, *Chem. Rev.* **109**, 5595–5619 (2009).

[10] K. Ravindranath. *Polymer* **31**, 2178–2184 (1990).

[11] P. J., J. Flory. *Am. Chem. Soc.* **58**, 1877–1885 (1936).

[12] R. Aris. *Introduction to the Analysis of Chemical Reactors*, (lst ed.). Prentice-Hall, Englewood Cliffs, NJ (1965).

[13] S. W. Benson. *Foundations of Chemical Kinetics*. (1st ed.). McGraw-Hill, New York (1960).

[14] E. Rabinowitch. *Trans. Faraday Soc.* 33, 1225–1233 (1937).
[15] P. J. Flory, *J. Am. Chern. Soc.* 61, 3334–3340 (1939).
[16] C. E. Carraher. *Seymour/Carraher's Polymer Chemistry*. CRC Press, Boca Raton, FL, (2008).
[17] W. A. Braunecker, and K. Matyjaszewski. *Prog. Polym. Sci.* **33**, 165 (2008).
[18] Y. Yoon, R. M. Ho, F. M. Li, M. E. Leland, J. Y. Park, S. Z. D. Cheng, V. Percec, P. W. Chu. *Prog. Polym. Sci.* **22**, 765 (1997).
[19] P. J. J. Flory. *Am. Chem. Soc.* **58**, 1877–1885 (1936).
[20] P. J. Flory. *Principles of Polymer Chemistry*, (1st ed.). Cornell University Press, Ithaca (1953).
[21] S. L. Rosen. *Fundamental Principles of Polymeric Materials.* Wiley, New York, NY (1982).
[22] C. C. Lin and K. H. Hsieh, T. *J. Polyrn. Sci.* **21**, 2711–2719 (1977).
[23] C. C. Lin and P. C. Yu. *Polym. Sci. Polym. Chem.* Ed. **16**, 1005–1016 (1978).
[24] C. C. Lin and P. C. Yu. *J. Appl. Polym. Sci.* **22**, 1797–1803 (1978).
[25] Ying-Hung So. *Acc. Chem. Res.* **34**, 753–758 (2001).

6

Chain Growth Polymerization

Chain growth polymerization is of great commercial importance. This polymerization commonly referred to as addition polymerization due to reaction occurs without any elimination of small molecule that occurs mainly in step growth polymerization. However, the polyaddition reactions take place in step growth polymerization, hence to differentiate addition without elimination of small molecule, this type of polymerization designated as chain growth polymerization. The way of growth of long chain molecules and conversion of monomer into polymer exploited widely in industry for the production of commercially important polymers such as polyethylene, polypropylene, etc. [1–5].

6.1 Monomer Type

Identical monomers undergo rapid reaction of addition with other monomer known as chain growth. Single chain growth, the length of polymer chain grows with an addition of one monomer. Based on different conditions, a large variety of unsaturated or cyclic compounds converted into polymer in chain growth polymerization. This polymerization results in linear polymers. Monomers of the general formula $H_2C = CRR'$ are readily polymerized after a proper initiation. R' is the hydrogen atom and then the resulting polymers are termed as vinyl polymers. Radical species activate monomer and propagates by activating neighboring monomers. Chain growth polymerization can be effectively polymerize the monomers with excellent yield. The monomer in the chain growth polymerization includes ethylene, styrene, and methyl methacrylate, etc. which are of particular industrial importance. The main types of chain growth monomers as listed in Table 6.1. Catalysts such as benzoyl peroxide, azobisisobutyronitrile, etc initiate monomers conversion to polymers. Chain growth polymerization is even at an early stage, the polymer

Table 6.1 List of commercially important monomers

Monomer	Chemical Structure
Ethylene	$CH_2=CH_2$
Propylene	$CH_2=CH-CH_3$
Vinylchloride	$CH_2=CHCL$
Styrene	$CH_2=CH-C_6H_5$
Methylmethacrylate	$CH_2=CCH_3COOCH_3$
Butadiene	$CH_2=CH-CH=CH_2$

formed has its final high molecular weight remaining the reaction mixture with unreacted monomer [6–8].

6.2 Reactions

Chain growth polymers are long chain molecules in which elementary building blocks linked together by covalent bonds. The reactions lead to the formation of commercially important polymers. Functionality of these monomer block links controls whether linear, branched or cross-linked macromolecules form. Three basic processes namely initiation, propagation, and chain transfer reactions to produce polymers.

Monomer addition and incorporation into the polymer happen from all atoms present in the monomer in polymerization. The functionality of these monomer block links controls whether linear, branched or cross-linked macromolecules form. The reactions lead to the formation of commercially important polymers. The addition occurs from monomers containing a double bond between carbon atoms. The growing chains at the reactive sites consume monomers.

Syndiotactic polymer over isotactic polymer configuration causes by steric and/or electric repulsion between the substituents present in the chain. These effects diminishes progressively the reaction at elevated temperatures.

6.2.1 Unsaturated Monomers

6.2.1.1 The polymerization of ethylene to polyethylene

Almost odorless and lighter gas of ethylene transforms into a hard and strong waxy solid polymer. The reaction goes in the formation of butane, hexane etc. until ultimately into polymer (Equations (1), (2) and (3)).

$$CH_2=CH_2 + CH_2=CH_2 \longrightarrow CH_2=CH\,CH_2CH_3 \quad \text{--------}\rightarrow (1)$$

$$CH_2{=}CHCH_2CH_3 + CH_2{=}CH_2 \longrightarrow CH_2{=}CHCH_2CH_2CH_2CH_3 \quad \text{--------}\rightarrow (2)$$

$$nCH_2{=}CH_2 \longrightarrow CH_2{=}CH(CH_2)_{2n-3}CH_3 \quad \text{--------}\rightarrow (3)$$

The catalyst in vinyl polymerization is extremely susceptible to catalytic action. As a rule, oxygen and peroxides are catalysts, however, antioxidants will inhibit polymerization. UV light stimulates the process. Polyvinylchloride forms simply by the polymerization of vinyl chloride (Equation (4)).

$$nCH_2{=}CHCl \longrightarrow CH_2{=}CH(\underset{Cl}{CH}CH_2)_nH \quad \text{--------}\rightarrow (4)$$

6.2.2 Monomer with End Groups

The reaction of monomer end groups of one to other in end group's addition reactions to produce new groups are the other type of chain growth polymerization. The reaction of isocyanate with an alcohol to form urethane (Equation (5)) is an example of these of chain growth polymers.

$$RNCO + HOR' \longrightarrow RNHCOOR' \quad \text{--------}\rightarrow (5)$$

The molecules grow into polymer rapidly once the reaction is initiated. The polymer grows to its final molecular weight, and then the reaction terminates. The purpose of a catalyst is to initiate the reaction by decomposing to generate reactive radical species. Successive addition of monomers and termination reactions occurs by mutual reaction, coupling, or disproportionation.

6.3 Polymerization

A radical may add to a double bond, generating a growth center at a new position. The radicals normally react further in an effort to stabilize itself, can add to another molecule and further addition is one of the mechanisms for chain growth polymerization. Three basic processes namely initiation, propagation, and chain transfer reactions to produce polymers [9–12].

Initiation, propagation, and termination reactions are three different kinds of chemical reactions with different mechanism and speed. A chain reaction in which the growth of a polymer chain proceeds exclusively by reaction(s) between monomer(s) and reactive site(s) on the polymer chain with regeneration of the reactive site(s) at the end of each growth step. Chain polymerization denotes chain reaction rather than polymer reaction.

6.4 Initiation

Chain initiation starts the chemical process by means of an initiator (Equations (6) and (7)). The typical initiators include any organic compound such as azo, disulphide, peroxide, etc. with a labile group. The commonly used peroxides are benzoyl peroxide and AIBN. The species are reactive species. The species are aggressive and difficult to control.

$$C_6H_5C(=O)O{-}OC(=O)C_6H_5 \longrightarrow C_6H_5^{\bullet} + CO_2 + C_6H_5C(=O)O^{\bullet} \quad \text{-------}\rightarrow (6)$$

Benzoyl peroxide — Phenyl radical — Benzoxy radical

$$(CH_3)_2C(CN){-}N{=}N{-}C(CN)(CH_3)_2 \longrightarrow (CH_3)_2\dot{C}(CN) + N_2 \quad \text{-------}\rightarrow (7)$$

Azobisisobutyronitrile — α−Cyanoisopropyl radical

$$R = C_6H_5^{\bullet} \text{ or } (CH_3)_2\dot{C}(CN)$$

Phenyl radical — α−Cyanoisopropyl radical

Radical (R°) can rapidly attack monomer. The product obtained must also be a radical which can in turn attack the monomer. Two growing species present even in exceedingly small concentration will find each other and react to give stable molecules by competing reaction in either coupling or disproportionation. Conversion of all monomers may require long time to become a polymer. Initiator generates radical species fragment, which gets incorporated as an end-group in the polymer chain [13–16].

In chain growth polymerizations, the growing radical species rapidly reacts with the C=C double bonds of the monomers (Equations (8) and (9)), which significantly exists in the reaction media, to yield the polymers efficiently. Two growing species present even in exceedingly small concentrations will find each other and react to give stable molecules by competing reaction in either coupling or disproportionation [17].

$$R^{\bullet} + {>}C{=}C{<} \longrightarrow R{-}C{-}C^{\bullet} \quad \text{-------}\rightarrow (8)$$

$$CH_2{=}CH(X) + R^{\bullet} \longrightarrow RCH_2{-}\dot{C}H(X) \quad \text{-------}\rightarrow (9)$$

6.5 Propagation

Radical formation may add to another double bond and continues as one of the mechanisms for chain growth polymerization. Propagation is far greater than termination rate. The chains grow by propagating these reactive sites through inclusion of monomers at such sites (Equations (10) and (11)). These inclusions are very rapid and chain-growth can take place in a fraction of a second, as the chains successively add monomers. Four propagation type of reactions are involved in the polymerization.

$$R-C-C^{\bullet} + C=C \longrightarrow R-C-C-C-C^{\bullet} \quad \text{-------}\rightarrow (10)$$

$$R-CH_2-\dot{C}H(X) + CH_2=CH(X) \longrightarrow R{-}[CH_2-CH(X)]{-}CH_2-\dot{C}H(X) \quad \text{-------}\rightarrow (11)$$

In propagation reactions, the free species either give stable product as termination reaction or lead to other species (Equations (12), (13) and (14)). The radical must usually react further as in the propagation reactions. Species formed from other species, addition either of peroxides or by the reaction between radical and a molecule or cleavage of species to give another radical or form by oxidation or reduction, which includes electrolytic methods [18–25].

- Abstraction of another atom usually hydrogen atom

$$R^{\bullet} + R-CH_2=CH(X) \longrightarrow R-H + R-\dot{C}H=CH(X) \quad \text{-------}\rightarrow (12)$$

- Rearrangement

$$R_3C-\dot{C}H_2 \longrightarrow R_2\dot{C}-CH_2R \quad \text{-------}\rightarrow (13)$$

- Decomposition

$$C_6H_5-C(=O)-O^{\bullet} \longrightarrow C_6H_5^{\bullet} + CO_2 \quad \text{-------}\rightarrow (14)$$

6.6 Termination Reactions

Reactions of free species reactions leading to stable products is termed as termination reaction. One of the termination reactions is a simple combination of similar or different species [26]. The lack of control in chain growth reactions are due to chain transfer and termination processes. This uncontrolled nature prevents synthesis of well-defined polymers with low polydispersity and complex architecture [27].

Chain termination occurs by either combination or disproportion. Side reactions such as chain transfer to monomer, solvent or polymer may occur. However, it also reacts rapidly with another growing radical chain end, which is present at a very low concentration, to result in the termination or abstracts hydrogen relatively slowly from C-H bonds in the polymers or solvents to induce chain transfer reaction. Radical coupling reaction, oligomerization, polymerization, etc., occur rapidly (Equations (15), (16) and (17)). Backbiting is a disadvantage of chain polymerization and normally side effects of the reaction (Equation (18)).

Coupling reaction or combination

$$R{+}\!\left[CH_2{-}\underset{X}{CH}\right]_n{-}CH_2{-}\underset{X}{\dot{C}H} + R{-}CH_2{-}\underset{X}{\dot{C}H} \longrightarrow R{+}\!\left[CH_2{-}\underset{X}{CH}\right]_{n+1}{-}R \quad \text{--------}\rightarrow (15)$$

Disproportionation

$$R\left[CH_2{-}\underset{X}{CH}\right]_n{-}CH_2{-}\underset{X}{\dot{C}H} \longrightarrow R\left[CH_2{-}\underset{X}{CH}\right]_n{-}CH{=}\underset{X}{CH} \quad \text{--------}\rightarrow (16)$$

$$\longrightarrow R\left[CH_2{-}\underset{X}{CH}\right]_n{-}CH_2{-}\underset{X}{CH_2} \quad \text{--------}\rightarrow (17)$$

Backbiting

$$R\left[CH_2{-}\underset{X}{CH}\right]_n{-}CH_2{-}\underset{X}{\dot{C}H} \longrightarrow \overset{X}{\dot{C}H}\left[CH_2{-}\overset{X}{\dot{C}H}\right]_n{-}CH_2{-}R \;;\; R\left[CH_2{-}\underset{X}{CH}\right]_n{-}CH{-}\underset{X}{CH_2} \quad \text{--------}\rightarrow (18)$$

6.7 Chain Growth Polymers

Chain growth polymers are obtained from monomers without any loss of small molecule. The polymer has the same composition as the monomer. Based on different conditions in chain growth polymerization, varieties of unsaturated or cyclic compounds get converted into polymers. Polymerization surpasses the reaction twice or thrice than the step growth polymerization Thus chain growth polymers result in linear polymers [28–31].

Chain growth polymers are practically compounds with double bonds. The reaction involves carbon-carbon double bonds. Chain growth polymers are long chain molecules in which elementary building blocks linked together by covalent bonds. Chain growth polymerization involves several stages to form a polymer. The first stage in chain polymerization is chain kinetic reaction.

6.7.1 Characteristics

Chain growth polymerization invariably proceeds by a chain reaction mechanism involving initiation, propagation, and termination. Monomer addition to the growing chain depends on the substituent, stability of the active center and steric hindrance for placing the approaching monomer to the growing chain. Most monomers which use chain growth polymerizations are nonsymmetrical.

The chain length formed during the early stages of polymerization is high. It will not be withstanding the gel or Trommsdorff effect. A reduction with conversion occurs due to the depletion of a monomer. The molecular weight distribution, polydispersity, and the ratio of weight/number and the average molecular weights are governed by a statistical factor. The relation between the degree of polymerization versus conversion in terms of monomer is represented in Figure 6.1.

The main characteristics of the polymerization are:

- Addition of a monomer repeating units are one at a time to the chain and an addition only to the active chain end makes the growth occur.

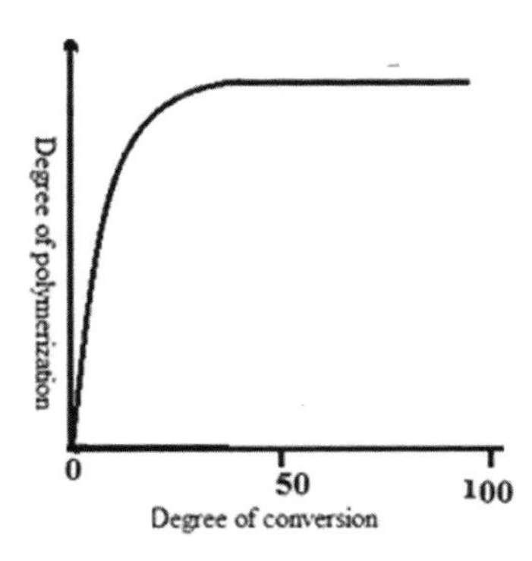

Figure 6.1 Relation between the degree of polymerization versus conversion – Chain growth polymerization.

- Monomer concentration decreases throughout the reaction in a steady manner.
- Molecular weight increases gradually with a formation of high molecular weight polymers.
- Long reaction time yields high productivity and affects the molecular weight.
- A monomer present with a decrease in the concentration and the impurities present will not affect the polymerization.
- Molecular weight and yield depend on the mechanism and only a monomer and a polymer are present during the reaction.
- Chain growth polymerization can occur even spontaneously.
- High molecular weight polymer formation occurs at low conversion rate.
- No small molecules, such as water, ammonia, etc. are eliminated.
- New monomer gets added up on the growing polymer chain through the reactive active center which can be a
    - free radical in free radical addition polymerization
    - carbocation in cationic addition polymerization
    - carbanion in anionic addition polymerization
    - organometallic complex in co-ordination polymerization
- The monomer molecule can be an unsaturated compound like ethylene or acetylene or a ring as in ring opening polymerization.

6.7.2 Classification

Substituents play a significant role in chain polymerization, because the polymerization ability of a vinyl monomer is dependent upon the steric and electronic properties of the substituents. The electronic effect of the substituent manifests itself by altering the electron density of the double bond through inductive and resonance effects and its ability to stabilize the active species, whether it is a radical, anion, or cation. Therefore, the chain growth polymerization can be classified [32–37] depend on the nature of the growth centers as

- Radical polymerization
- Cationic polymerization
- Anionic polymerization
- Stereoregular polymerization

6.7.3 Growth Centers

In chain growth polymerization, there are growth centers in the reaction. In the growth centers, the monomers add on successively. The repeating unit of a chain growth has the same composition as the monomer. From the beginning, the reaction produces high molecular weight product. The unreacted monomer quantity decreases with time. The chain growth polymerization is further classified based on the nature of the growth centers. They are radical, cationic, anionic, and stereoregular polymerization reactions [38–45].

6.7.4 Structural Variations

Chain growth polymerizations are non-symmetrical results in three possible sequences of the monomer units in the chain. These include head – to – tail, tail – to – tail and head – to – head structures termed as sequence isomerism. In some cases, random grouping may also occur.

Structural variations in chain growth polymers are

1. Structural isomerism due to the formation of branches in homopolymers and variation in the monomer distribution in copolymers
2. Sequence isomerism arising from variations in orientation of asymmetric monomer units
3. Stereoisomerism arising from differences in configuration of asymmetric C atoms in the chain and
4. Geometric isomerism in diene polymers cis and trans.

6.8 Kinetics

Reaction is composed of three different reactions namely initiation, propagation, and termination. Initiator generates radical fragment that get incorporated as an end-group in the polymer chain.

$$\text{C=C} \xrightarrow[\text{double-bond reaction}]{\text{chain polymn}} \text{-[C-C]}_n\text{- Vinyl polymer} \quad \text{---------} \rightarrow (19)$$

Chain polymerization is the simplest and most frequently used polymerization technique (Equation (19)). CH_2–CHR, a vinyl monomer forms polymer –[–H_2C–CH(R) –]$_n$. Chain growth polymerization is a double bond cleavage reactions. Molecular weight of the polymer and the end group content is useful to establish the ratio of termination by coupling versus disproportionation. The schematic representation of chain growth polymerization is in Figure 6.2.

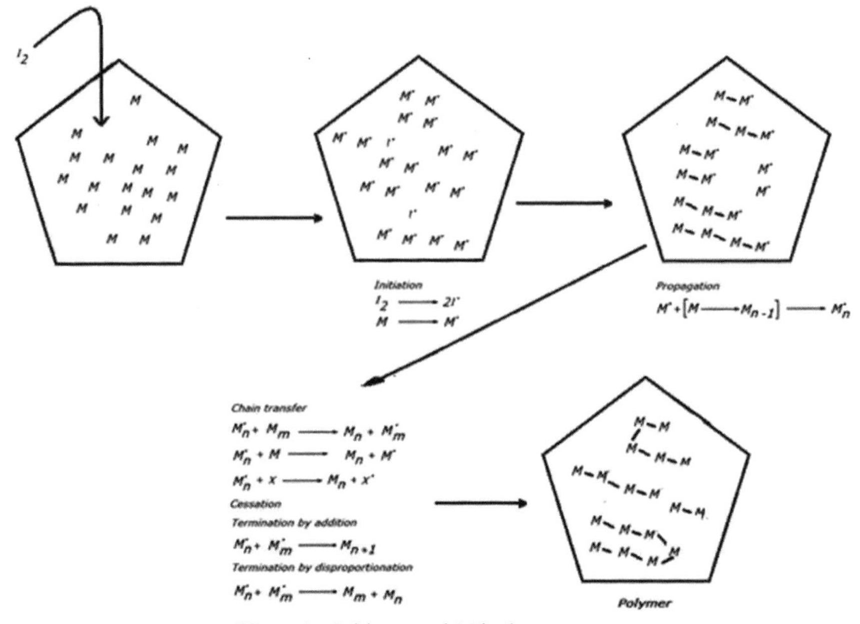

Where I = Initiator and M is the monomer.

Figure 6.2 Chain growth polymerization.

Initiation, propagation, and termination reactions are three different kinds of chemical reactions with different mechanism and speed. A chain reaction in which the growth of a polymer chain proceeds exclusively by reaction(s) between monomer(s) and reactive site(s) on the polymer chain with regeneration of the reactive site(s) at the end of each growth step. Chain polymerization denotes chain reaction rather than polymer reaction. Two growing species present even in exceedingly small concentration will find each other and react to give stable molecules by competing in reaction for either coupling or disproportionation.

Chain polymerization requires an initiator (I) and proceeds by adding one repeating unit at a time is illustrated below [46]:

$$
\begin{aligned}
I+M &\longrightarrow R_1 \\
M+R_1 &\longrightarrow R_2 \\
M+R_2 &\longrightarrow R_3 \\
M+R_3 &\longrightarrow R_4, \text{ etc.}
\end{aligned}
$$

The molecular weight rapidly builds once chain initiates.

The basic steps in free radical polymerization are initiation, propagation, chain transfer, and termination.

6.8.1 Initiation

In chain polymerization, an initiation step is needed to initiate the polymer chain growth. Initiation can be achieved by adding chemicals that decompose to form free species. Small amount of addition of initiators can be mono-functional and form the same species (I).

$$I_2 \xrightarrow{k_o} 2I$$

The reaction between monomer M and initiator I for mono-functional initiators as follows:

$$I + M \xrightarrow{k_i} R_i$$

The rate of initiation of the formation of initiator (I) from the chemical (I_2) as in equation [47]. There will be scavenging or recombining of the primary species, only a certain fraction f will be in initiated as primary species. Each reaction step is assumed to be elementary. Therefore, the rate law for the formation of the initiator species r_i *is*

$$-r_{I_2} = k_o I_2$$
$$-r_i = k_i MI$$

6.8.2 Propagation

Specific reaction rate k_i is often taken to be equal to k_p [48].

k_p is assumed to be identical for the addition of each monomer to growing chain.

$$R_p + M \xrightarrow{k_p} R_{p+1}$$

The rate of propagation as follows

$$-r_p = k_p MR_p$$

6.8.3 Termination

In chain transfer reactions, the species involved in the chain transfer reactions are all assumed to have the same reactivity immaterial of the species size as R_i. In other words, R_i is produced in chain transfer reactions are taken to be the same. However in some cases, the chain transfer agent may be too large or unreactive to propagate the chain. The choice of solvent is important, the solvent transfer reaction rates k_s is greater between two different solvents.

The specific reaction rates in chain transfer are all assumed to be independent of the length of the chain. The species produced R_i in each of the chain transfer steps are different. The reaction is essentially the same manner as the species R_i in the propagation step to form another species R_p.

Molecular weight of the polymer and the end group content is useful to establish the ratio of termination by coupling versus disproportionation. Coupling produces polymer with and disproportionation produces only one radical fragment per polymer molecule.

6.8.3.1 Chain transfer reactions

Transfer reactions play an important role in the kinetics of the reaction. Completion of the reaction takes place within a few seconds. Transfer reactions can be considered cessation reactions. They terminate the growth of any particular chain with no effect on the overall rate of polymerization because a new active nucleus is produced as each growing chain is terminated. They may play an important role in the formation of branched chains.

The formation of polymer contains thousands of monomer. Coupling produces polymer with and disproportionation produces only one radical fragment per polymer molecule.

The transfer of species from a growing polymer chain can occur as mentioned below:

- Transfer to a monomer – a live polymer chain of p monomer units transfers its species to the monomer to form R_i and a polymer chain of m monomer units.

$$R_p + M \xrightarrow{k_m} P_m + R_i$$

The rate of law of transfer to a monomer is

$$-r_{mp} = k_m M R_p$$

- Transfer with another species

$$R_p + C \xrightarrow{k_c} P_m + R_i$$

The rate of law of transfer with another species is

$$-r_{cp} = k_c C R_p$$

- Transfer to solvent

$$R_p + S \xrightarrow{k_s} P_m + R_i$$

The rate of law of transfer of solvent to growing polymer is

$$-r_{sp} = k_s S R_p$$

Termination occurs primarily by two mechanisms to form polymer.

- Termination by addition of two growing polymers

$$R_p + R_g \xrightarrow{k_{ta}} P_{m+g}$$

Rate of law of addition of two growing polymers is

$$-r_{ap} = k_{ta} R_p R_g$$

- Termination by disproportionation

$$R_p + R_g \xrightarrow{k_{td}} P_m + P_g$$

The rate law of termination by disproportionation is

$$-r_{dp} = k_{td} R_p R_g$$

6.9 Non-Biodegradability

The non-biodegradable nature of chain growth of polymers are generally based on

- Take millions of years to degrade due to strong covalent bond linkage between monomers.
- Molecular weight is higher.
- Hydrophobicity nature is higher.

6.10 Summary

- In chain growth polymerization, reactions take place with repeated addition of monomers.
- Monomer consumption proceeds with growing chains at the reactive sites.
- Polymerization may be initiated-thermally, through free radicals, or ions.
- The chain polymerization invariably proceeds by a chain-reaction mechanism involving three elementary steps, i.e. initiation, propagation, and termination.
- The chain growth polymerization is the choice for many vinyl monomers which are available commercially at low cost.
- The method of initiation may be of great importance in determining the physical properties of the product, but it is of secondary importance in its effect on reactor behaviour.

References

[1] D. H. Solomon, E. Rizzardo, and P. Cacioli. *Eur. Patent Appl.* EP135280; *Chem. Abstr.*, **102**, 221335 (1985).
[2] D. H. Solomon, E. Rizzardo, and P. Cacioli. *U.S. Patent 4, 581, 429*, March 27 (1985).
[3] E. Rizzardo. *Chem. Aust.*, **54**, 32 (1987).
[4] C. H. L. Johnson, G. Moad, D. H. Solomon, T. Spurling, and D. J. Uearing. *Aust. J. Chem.*, **43**, 1215 (1990).
[5] G. Moad, and E. Rizzardo. *Macromolecules*, **28**, 8722 (1995).
[6] M. K. Georges, R. P. N. Veregin, P. M. Kazmaier, and G. K. Hamer. *Macromolecules*, **26**, 2987 (1993).
[7] M. K. Georges, R. P. N. Veregin, P. M. Kazmaier, and G. K. Hamer. *Trends Polym. Sci.*, **2**, 66 (1994).
[8] M. K. Georges, R. P. N. Veregin, P. M. Kazmaier, and G. K. Hamer, M. D. Saban. *Macromolecules*, **27**, 7228 (1994).
[9] I. Q. Li, B. A. Howell, K. Matyjaszewski, T. Shigemoto, P. B. Smith, and D. B. Priddy. *Macromolecules*, **28**, 6692 (1995).
[10] I. Q. Li, B. A. Howell, R. A. Koster, and D. B. Priddy. *Macromolecules*, **29**, 8554 (1996).

[11] I. Q. Li, B. A. Howell, M. T. Dineen, P. E. Kastl, J. W. Lyous, D. M. Meunier, P. B. Smith, and D. B. Priddy. *Macromolecules*, **30**, 5195 (1997).
[12] S. Gaynor, D. Greszta, D. Mardare, M. Teodorescu, and K. J. Matyjaszewski. *Macromol. Sci., Pure Appl. Chem.*, **A31**, 1561 (1994).
[13] P. G. Odell, R. P. N. Veregin, L. K. Michalak, D. Brousmiche, and M. K. Georges. *Macromolecules*, **28**, 8453 (1995).
[14] R. P. N. Veregin, P. G. Odel, L. M. Michalak, and M. K. Georges. *Macromolecules*, **29**, 3346 (1996).
[15] R. P. N. Veregin, P. G. Odell, L. K. Michalak, and M. K. Georges. *Macromolecules*, **29**, 4161 (1996).
[16] N. A. Listigovers, M. K. Georges, P. G. Odell, and B. Keoshkerian. *Macromolecules*, **29**, 8992 (1996).
[17] G. Moad, and D. H. Solomon. *The Chemistry of Radical Polymerization*, 2nd ed. Elsevier, Oxford, 1–9 (2006).
[18] E. Yoshida, and T. J. Fujii. *Polym. Sci., Part A: Polym. Chem.*, **35**, 2371 (1997).
[19] E. Yoshida, and S. Tanimoto. *Macromolecules*, **30**, 4018 (1997).
[20] M. Steenbock, M. Klapper, K. Mullen, and M. Pinhal. *Acta Polym.*, **47**, 276 (1996).
[21] D. Bertin, and B. Boutevin. *Polym. Bull.*, **37**, 337 (1996).
[22] B. Améduri, B. Boutevin, and P. Gramain. *Adv. Polym. Sci.*, **127**, 87 (1997).
[23] G. Schmidt-Naake, and S. Butz. *Macromol. Rapid Commun.*, **17**, 661 (1996).
[24] Y. Yagci, A. B. Duz, and A. Onen. *Polymer*, **38**, 2861 (1997).
[25] M. Baumert, and R. Mülhaupt. *Macromol. Rapid Commun.*, **18**, 787 (1997).
[26] K. Matyjaszewski, *Controlled Radical Polymerization; ACS Symposium Series 685*. American Chemical Society, Washington, DC (1998).
[27] M. Stickler, D. Panke, and A. E. Hamielec. *J. Polym. Sci., Polym. Chem. Ed.* **22**, 2243 (1984).
[28] C. J. Hawker, and J. L. Hedrick. *Macromolecules*, **28**, 2993 (1995).
[29] C. J. Hawker, E. Elce, J. Dao, W. Volksen, T. P. Russell, and G. G. Barclay. *Macromolecules*, **29**, 2686 (1996).
[30] C. J. Hawker. *Trends Polym. Sci.*, **4**, 183 (1996).
[31] C. J. Hawker, G. G. Barclay, A. Orellana, J. Dao, and W. Devonport. *Macromolecules*, **29**, 5245 (1996).

[32] O. Nuyken, and G. Lattermann. *Handbook of Polymer Synthesis Part A*, 1st ed, (Kricheldorf, H. R., ed.), Marcel Dekker, New York, 223ff (1991).
[33] R. Fittig, and L. Paul. *Justus Liebigs Ann. Chem.*, **188** 55 (1887).
[34] R. Fittig, and F. R. Engelhorn. *Justus Liebigs Ann. Chem.*, **200** 70 (1880).
[35] G. W. A. Kahlbaum. *Ber. Dtsch. Chem. Ges.*, **13** 2348 (1880).
[36] A. L. Barker, and G. S. Skinner. *J. Am. Chem. Soc.*, **46** 403 (1924).
[37] E. H. Riddle. *Monomeric Acrylic Esters*. Reinhold, New York (1954).
[38] P. J. Flory. *Principles of Polymer Chemistry*, 1st ed., Cornell University Press, Ithaca, New York (1953).
[39] R. W. Lenz. *Organic Chemistry of Synthetic High Polymers*, 1st ed., Wiley, New York (1967).
[40] A. Kumar and S. K. Gupta. *Fundamentals of Polymer Science and Engineering*, 1st ed., Tata McGraw-Hill, New Delhi (1978).
[41] P. E. M. Allen and C. R. Patrick. *Kinetics and Mechanisms of Polymerization Reactions*, 1st ed., Ellis Horwood, Chichester (1974).
[42] J. Furukawa and O. Vogl. *Ionic Polymerization, Unsolved Problems*, 1st ed., Marcel Dekker, New York (1976).
[43] T. Keii. *Kinetics of Ziegler-Natta Polymerization*, 1st ed., Kodansha, Tokyo (1972).
[44] J. Boor. *Ziegler-Natta Catalysts and Polymerizations*, 1st ed., Academic, New York (1979).
[45] J. C. W. Chien. *Coordination Polymerization*, 1st ed., Academic, New York (1975).
[46] N. A. Dotson, R. Galvin, R. L. Lawrence, and M. Tirrell. *Polymerization Process Modelling*, VCH Publishers, New York (1996).
[47] D. C. Timm, and J. W. Rachow. *ACS Symp. Ser.* **133**, 122 (1974).
[48] J. J. Kiu, and K. Y. Choi. *Chem. Eng. Sci.* **43**, 65 (1988); K. Y. Choi, and G. D. Lei. *AIChE J.* **33**, 2067 (1987).

7

Polymerization Reactions

Polymer chemistry has played a central role in shaping useful products. It reaches its potential only after it undergoes analysis, fractionation, and transformation. It is the most expressive in view of its creativity and unlimited scope. It has become a mature science with all the advantages and with the handicaps of maturity.

Polymer chemistry of polymers containing covalent bonds of monomers units are connected with each other. Polymers are separate from monomers and are a very common class of macromolecular compounds. Moreover, the toolbox of polymer chemistry contains polymerization methods that allow synthesis of polymers. Therefore, the chemical industry converted many of the monomers into polymers.

Direct polymerization of monomers is a more attractive strategy. Apart from the fact that the polymerization with functional groups strategy is highly attractive to generate a polymer; this facilitates the establishment of structure-property relationships to a great extent [1–4]. However, the polymers with precise control over molecular weight, composition and architecture still pose a significant challenge. There is still a broad range of polymers that have not been introduced by direct polymerization. Most of the knowledge of polymerization reactions and polymer properties comes from synthesis and experiments. Polymerization reactions prefer the media for many industrial and laboratory polymerization processes [5–8]. New monomer adds on the growing polymer chain through the reactive active center which can be a:

- Free radical in radical polymerization
- Carbocation in cationic polymerization
- Carbanion in anionic polymerization
- Organometallic complex in co-ordination polymerization.

7.1 Chemistry of Radicals

Radicals are reactive species, which undergo radical coupling reaction, oligomerization, polymerization, etc. occur rapidly, and control of reactions is not easy. Mild and excellent radical reactions established the fundamentals of organic free radicals. Polymerization with radical is a chain reaction [9].

7.1.1 Types of Radicals

Most organic radicals are quite unstable and very reactive. There are two kinds of radicals, neutral radicals and charged radicals as shown below, i.e. a neutral radical, is a cation radical and an anion radical (Figure 7.1)

There are σ radicals and π radicals. The σ radicals with an unpaired electron is in the σ orbital such as phenyl radical, vinyl radical, etc. π radicals appear with an unpaired electron in the π orbital such as tert-butyl radical. π radicals are stabilized by hyperconjugation or resonance effect. However, σ radicals are very reactive because there is no such stabilizing effect to withhold radical by some effect.

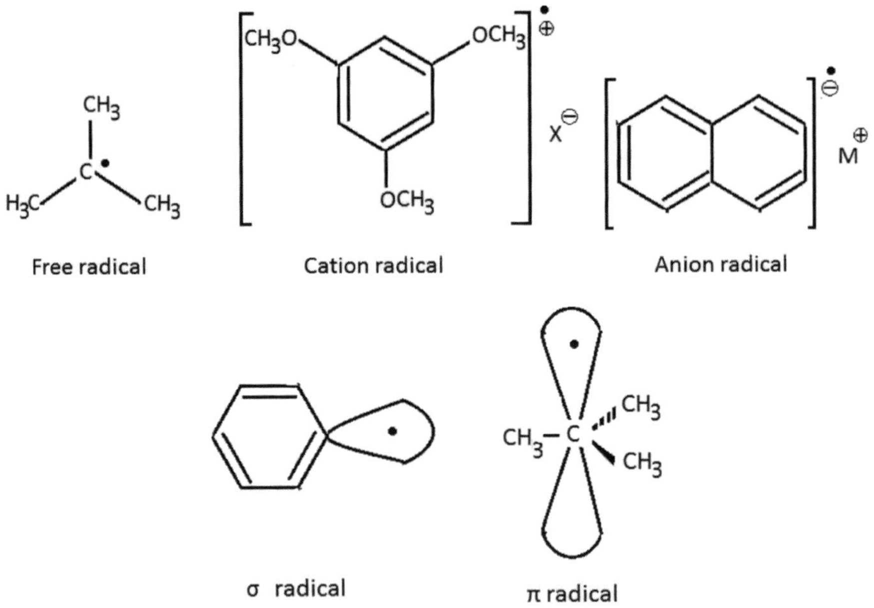

Free radical

Cation radical

Anion radical

σ radical

π radical

Figure 7.1 Neutral and charged radicals.

7.2 Radical Polymerization

Radical polymerization is an integral part of polymer chemistry [10]. In radical polymerization, a balance of several elementary reactions such as initiation, propagation; termination and chain transfer dominate happen. Steric hindrance leads to an increase in the polymerization reactivity. Steric effects take more important role in termination than in propagation. Many polymers are prepared by radical process involving high pressure, high temperature and a catalyst is usually a substance (organic peroxide), which readily breaks up to form radicals, which in turn initiates a chain reaction. The polymerizations allow only a limited degree of control [11]. The reactivity ratios of monomers and rates of polymerization controlled by the addition of Lewis acids [12–14].

Radical is simply a molecule with an unpaired electron. Initiator creates free radicals which forms two fragments along a single bond. Typical initiators include any organic compound such as azo, disulphide, peroxide, etc. with a labile group. Commonly used peroxides are benzoyl peroxide and AIBN. Initiator for radical polymerization includes any organic compound with a labile bond, such as an azo (- N – N-), disulfide (-S – S -), or peroxide (- O – O -) compound. Either by heat or irradiation such as UV or γ radiation giving a radical. The initiation takes place by cleavage to yield the "primary radical". Generally, the initiation takes place by cleavage of an azo or peroxide compound (the "initiator") to yield the "primary radical".

R – O – O – R is an initiator where R represents a benzoyl group (Equations (1) and (2)). The formation of a radical from benzoyl peroxide is initiator as given below:

$$C_6H_5\text{-}C(=O)\text{-}O\text{-}O\text{-}C(=O)\text{-}C_6H_5 \longrightarrow 2\ C_6H_5\text{-}C(=O)\text{-}O^{\bullet} \quad (1)$$

Benzoyl free radical
Active center

$$C_6H_5\text{-}C(=O)\text{-}O^{\bullet} \longrightarrow C_6H_5^{\bullet} + O{=}C{=}O \quad (2)$$

The stability of a radical refers to the tendency of a molecule to react with other compounds. An unstable radical will readily combine with many different molecules. However, a stable radical will not easily interact with other chemical substances. The stability of free radicals can vary widely depending on the properties of the molecule.

The increase of the solvent polarity results in an increase of the reaction rate in the polymerization of monomers such as ethers of methacrylic or acrylic acids [15, 16]. Radical polymerization is a chain growth polymerization. Propagation involves sequential addition of monomer. The formation continues to form propagating radicals.

7.3 Reaction Process

The tendency for the radical to gain an additional electron in order to form a pair makes it highly reactive so that it breaks the bond on another molecule by stealing an electron, leaving that molecule with an unpaired electron, which is another free radical. One of the two π electrons in the ethylene used to form a single bond with the R – O° radical. The other remains on the second carbon atom, leaving it as a seven – valence electron atom that reacts with another monomer. Not all chains can be simultaneously active.

Radical polymerization is a chain reaction and starts with radical formation from an initiator adds to the monomer during polymerization reaction that initiate monomers into chains. Radical polymerizations are composed of three primary processes initiation, propagation, and termination.

Radical polymerization consists of elementary reaction as:

1. Initiation reactions, which continuously generate radicals derived from the initiator called a primary or initiator radical during polymerization. For a radical polymerization – Chain initiation, by means of an initiator that starts the chemical process [9].

   Radicals initiate the chains formed from an initiator adding to monomer (Equation (3)). Involvement of sequential addition to monomer, the chain propagates. The propagating radicals terminate the reaction self-react by either combination or disproportionation. Addition can occur at either end of the monomer. The radical initiator attacks and attaches to a monomer. Activated monomer is a new free radical.

$$\underset{\text{Free radical}}{RO^{\bullet}} + \underset{\text{monomer}}{CH_2{=}CHX} \longrightarrow \underset{\text{Activated monomer}}{RO{-}CHX{-}CH_2^{\bullet}} \quad \text{- - - - - - - - - - - -}\rightarrow \quad (3)$$

Where $RO^{\bullet} = C_6H_5{-}C(=O){-}O^{\bullet}$ and $X = H, Cl$, etc.

One of the two π electrons in the monomer used to form a single bond with the R – O° radical. The other remains on the second carbon atom, leaving it as a seven – valence electron atom that reacts with another ethylene monomer.

2. Chain propagation reactions, which are responsible for the growth of polymer chains by addition of a monomer to a radical center.
3. In the propagation phase, the newly activated monomer attacks and attaches to the double bond of another monomer molecule. The process of electron transfer and consequent motion of the active center down the chain proceeds. The active center is the location of the unpaired electron on the radical, where the reaction takes place.

$$\sim\!\!\sim CHX-CH_2^{\bullet} + CH_2{=}CHX \longrightarrow \sim\!\!\sim CHX-CH_2-CHX-CH_2^{\bullet} \quad \text{- - - - - - - - - - -} \rightarrow \quad (4)$$

In radical polymerization, the entire propagation reaction (Equation (4)) usually takes place within a fraction of second. Addition of thousands of monomers to the chain take place within the time. The entire process steps when the termination reaction occurs. Propagation involves sequential addition of monomer to form propagating radicals. This formation continues to from propagating radicals. The propagation reaction in radical polymerizations is rapid [17–22]. The chain length formed during the early stages of polymerization is high. It will not be withstanding the gel or Trommsdorff effect. Reduction with conversion occurs due to the depletion of monomer.

In theory, the propagation reaction (Equation (5)) could continue until the supply of monomers is exhausted. However, most often the growth of a polymer chain halted by termination reaction. The chain reaction must terminate. Termination typically occurs by combination and disproportionation. Combination occurs once polymer chain growth stops. With radical end groups on the two growing chains that join and form a single chain.

$$\sim\!\!\sim CHX-CH_2-CHX-CH_2^{\bullet} + {}^{\bullet}CH_2-CHX-CH_2-CHX\sim\!\!\sim \longrightarrow$$
$$\sim\!\!\sim CHX-CH_2-CHX-CH_2-CH_2-CHX-CH_2-CHX\sim\!\!\sim$$
$$\text{- - - - - - - - - - -} \rightarrow \quad (5)$$

4. Chain termination occurs by either combination or disproportion. Side reactions such as chain transfer to monomer, solvent or polymer may

occur. Termination occurs by self-react by combination or disproportionation. Not all chains can be simultaneously active. The propagation or growth reaction occurs when monomers add to the primary radical or to the radical at the end of the growing polymer chain.

- Bimolecular termination reactions (Equation (6)) between two radical centers, which give a net consumption of radicals. These consist of disproportionation and combination.

  Disproportionation halts the propagation reaction when a radical strips a hydrogen atom from an active chain. Carbon-carbon double bond takes the place of the missing hydrogen. This can also occur when the radical reacts with an impurity. Therefore, it is important that polymerization should be carried out under very clean conditions.

$$\sim\sim CHX-CH_2-\dot{C}HX-\dot{C}H_2 + {}^{\bullet}CH_2-CHX-CH_2-CHX\sim\sim \longrightarrow$$

$$\sim\sim CHX-CH_2-CHX-CH_3 + CH_2{=}CHX-CH_2-CHX\sim\sim \quad (6)$$

The polymerization can be controlled the way a polymer does each of these steps by varying the reactants, the reaction times and the reaction condition. The physical properties of a polymer chain depend on the polymer average length, the amount of branching and the constituent monomers. Backbiting (Equation (7)) is an unwanted reaction during polymerization that creates side chains.

$$-\underset{H}{\overset{H}{C}}-CHX-CH_2-CHX-\dot{C}H_2 + \dot{C}H_2{=}CHX \longrightarrow -\underset{H}{\dot{C}}-CHX-CH_2-CHX-CH_2-CHX-CH_3 \quad (7)$$

Termination occurs primarily by the bimolecular reaction of two growing polymer radicals. There are radical-radical combination and disproportionation in which one radical abstracts a hydrogen atom from another radical resulting in one saturated chain end and one double bond chain end. There may be very low levels of chain transfer to monomer, a reaction that leads to termination of one chain with simultaneous initiation of another new chain so that there is no change in the number of radical species present. The concentration of entanglement points increases during the course of polymerization. High monomer conversion, the polymer radicals trapped. Therefore, radical centers may continue to move and undergo bimolecular termination [23]. In case of low monomer conversion, polymer chains in a good solvent are isolated coils, translational diffusion of the centre of the mass of the chains is sufficiently rapid.

- Chain transfer to small molecules that causes the cessation of growth of polymer radicals while generating small transfer radicals simultaneously. Radicals would recombine too quickly with another radical, either through disproportionation or through a coupling reaction. The disproportionation reactions will produce both saturated and unsaturated chain end. Radicals recombine and disproportionate with rates close to the diffusion controlled limits.

Polymerization is subject to many complications. However, with simple mechanism, the unsaturated intermediates that capable of adding themselves during reaction [24, 25]. In general, radical reactions are not selective. The polymerizations allow only a limited degree of control. The use of chain terminators employed to control molecular weights [26–30]. Radical polymerization finds application in the synthesis of many important classes of polymers including those based upon methacrylates, styrene, chloroprene, acrylonitrile, ethylene, and the many copolymers of these vinyl monomers.

7.3.1 Advantages

- A majority of commercial polymerizations are radical polymerizations, which are widely used processes for the commercial production of high-molecular weight polymers.
- The main factors responsible for the preeminent position of radical polymerization are the ability to polymerize a wide array of monomers including (meth)acrylates, (meth)acrylamides, acrylonitrile, styrene, dienes, and vinyl monomers and compatibility with reaction conditions.
- High molecular weight polymer forms at low conversion rate.
- Elimination of no small molecules, such as water, ammonia, etc.
- Radical polymerization has simple conditioning, easy processing, applicability to aqueous system, etc.

- The tolerance of unprotected functionality is in monomer, solvent, and compatibility with a variety of reaction conditions.
- Tolerance of unprotected functionality in monomer and solvent (e.g., OH, NR_2, COOH, $CONR_2$, and SO_3H) (polymerizations can be carried out in aqueous or protic media).
- Simple to implement and inexpensive in relation to competitive technologies with widely used industrial method due to generation of a radical is easy that can initiate polymerization.

7.3.2 Disadvantages

- The polymerization itself is difficult to be controlled due to continuous generation of radicals.
- Heat generation occurs due to exothermic reaction.
- Severely limits the degree of control that can assert over molecular-weight distribution, copolymer composition, and macromolecular architecture [31–34].
- Randomly occurring termination reaction between radicals coming by each other. This termination made it impossible to get polymers with monodispersed chain lengths and controlled molecular architecture.
- Yield polymers with uncontrolled molecular weights and high polydispersity, which prevents the synthesis of well-defined polymers with low polydispersity and complex architectures.

7.4 Ionic Polymerization

The selectivity of ionic polymerization stabilization of anionic and cationic propagation species in solvents with high solvating power where the distance between the propagating active center and the counter ion is large, the factors governing the stereochemistry are similar to those for radical polymerization. While polar and highly solvating media for ionic polymerizations many not be used due to reaction and negate the ionic initiators.

Ion initiated polymerization follows the same basic steps like free radical polymerization (initiation, propagation, and termination). However, there are some differences present. Ionic mechanism of chain polymerization involves attack on the π electron pair of the monomer.

Either in ionic polymerization, carbonium (C^+) (Equation (8)) or a carbanion (C^-) (Equation (9)) ionic site forms in the initiation process. Radical – initiated polymerization are generally non-specific but this is not true for ionic initiators, since the formation and stabilization of a carbonium ion or carbanion depends largely on the nature of the group X and Y in the vinyl monomer.

$$CH_2{=}CHX \longrightarrow CH_2{-}\overset{\oplus}{C}H \xrightarrow{CH_2=CHX} -CH_2{-}\underset{X}{\underset{|}{C}H}{-}CH_2{-}\overset{\oplus}{C}H \dashrightarrow \quad (8)$$

The X has an electron-donating group such as alkoxy, phenyl, vinyl, etc., on to stabilize the cationic intermediate.

$$CH_2{=}CHY \longrightarrow CH_2{-}\overset{\ominus}{C}H \xrightarrow{CH_2=CHY} -CH_2{-}\underset{Y}{\underset{|}{C}H}{-}CH_2{-}\overset{\ominus}{C}H \dashrightarrow \quad (9)$$

The Y has an electron-withdrawing group such as – CN, CO, Phenyl, vinyl, etc., on to stabilize the anionic intermediate. Cationic and anionic polymerizations have many similarities in character. Stabilization of the propagating centers by solvation generally requires yield of high molecular weight polymers. To suppress termination, low or moderate temperatures are needed and other transfer and chain-braking reactions that destroy propagating centers.

Solvents of high polarity are desirable to solvate the ions. The polar hydroxyl solvents such as water, alcohols, etc., react and destroy many of the ionic initiators. Polar solvents such as ketones prevent initiation of polymerization by forming highly stable complexes with the initiators. Ionic polymerizations usually carried out in solvents of low or moderate polarity.

When the chain carrier is ionic, the reaction rates are rapid, difficult to reproduce, and yield high molar mass material at low temperatures by mechanisms, which are often difficult to define. Initiation of an ionic polymerization are much more complex than radical reactions. Initiation of an ionic polymerization can occur in one of four ways involving essentially the loss or gain of an electron by the monomer to produce an ion or radical ion.

Bimolecular termination between active centers does not occur in ionic polymerization. Termination of an active center on a polymer chain occurs by reaction with the counterion, solvent, monomer, or other species. Often, the initiation reactions are very fast, and the initiator consumed in the early stages of polymerization before the polymer chains have grown much beyond oligomeric size. In the absence of uni-molecular termination, a population of polymer chains having the same molecular mass can grow. The concentration of ionic reactive centers is usually much larger than that of radical centers. Ionic polymerization have requirements of extreme purity for the components of the reacting mixture [35–37].

Ionic polymerizations at low pressures, the polymerization may proceed through elimination reactions. At high pressure, the ionic intermediates stabilized and addition without elimination may occur.

7.5 Types of Anions

Negatively charged organic species in which carbon carries three bond pairs and one lone pair are commonly called as carbanion (Equations (10) and (11)). It has a more powerful base and a stronger nucleophile in polymerization reactions. It is sufficiently basic to remove a proton from the monomer. It represents as

$$M + I^{-} \longrightarrow MI^{-} \qquad (10)$$

$$M + e^{-} \longrightarrow M^{-} \qquad (11)$$

Where M = monomer and I^{-} is the carbanion.

Anionic polymerization is initiated by anion may be base or nucleophile such as n-butyl lithium or potassium amide. Monomer containing electron withdrawing groups like phenyl, nitrile, etc undergo this polymerization.

7.6 Types of Cations

Cationic polymerization is initiated by an acids with generation of carbonium ions (Equations (12) and (13)) commonly Lewis acids such as BF_3, $FeCl_3$, $AlCl_3$, $SnCl_4$, H_2SO_4, HF in presence of small amount of water.

$$M + I^{+} \longrightarrow MI^{+} \qquad (12)$$

$$M - e^{-} \longrightarrow M^{+} \qquad (13)$$

Cationic initiation is usually limited to monomers with electron-donating groups such as alkoxy, alkenyl, phenyl, etc. which help to stabilize the delocalization of the positive charge in the π orbitals of the double bond. Ionic initiated polymerizations are much more complex than radical reactions.

7.6.1 Advantages

- Precise control of molecular weight and its distribution, end groups, pendant groups, sequence of distribution and steric structure of the polymer.
- Well-defined architectural polymers such as star shaped polymer becomes accessible with control.
- Initiation must be equal to or faster than propagation to ensure that all of the polymer chains start at the same time.

7.6.2 Disadvantages

- Ionic polymerizations are highly selective due to the very strict requirements.
- Utilization of cationic and anionic polymerization is rather limited due to high selectivity of ionic polymerization compared to radical polymerizations.
- The nature of the reaction in ionic polymerization is not clear due to involvement of heterogeneous inorganic initiators.
- Ionic polymerization proceeds at rapid rates and extremely sensitive to the presence of small concentrations of impurities and other ionic reactive species.
- Extreme dependence on and susceptibility of ions to impurities in the solvents and monomers triggers look like inconsistent results.
- Nonpolar solvents, such as hexane, have the opposite effect.
- Impurities such as moisture, protic solvents, etc. that normally would terminate the polymerization.

7.7 Cationic Polymerization

Cationic polymerization has rejuvenated the field of polymerization by creating new opportunities for polymer synthesis. Cationic polymerization is initiated by proton or carbocation derived from a strong super acid or an initiator-Lewis acid pair respectively. Small oligomers accumulate at the interface at the beginning of the reaction and decrease with depletion of water content at the interface. The effect is responsible for the increase in molecular weight with increasing monomer conversion.

It undergoes living polymerization by a variety of initiator – co-initiator pairs. Preparation of advanced materials is possible and not possible with other conventional polymerizations [38–40]. Strong protic acids used to from cationic initiating species (Equation (14)). High concentrations of the acid needed in order to produce sufficient quantities of the cationic species. The counter ion (A^-) produced must be weakly nucleophilic so as to prevent early termination due to combination with the protonated olefin (Equation (15)). Common acids such as phosphoric, sulfuric, fluro, and trific acids are used. Low molecular weight polymers formed with these acids.

$$HA \rightleftharpoons H^{\oplus}A^{\ominus} \quad \text{-------------} \rightarrow (14)$$

$$H^{\oplus} + CH_2{=}CH{-}R \longrightarrow CH_2{-}\overset{\ominus}{C}H{-}R \quad \text{-------------} \rightarrow (15)$$

Cationic polymerization proceeds through attack on the monomer by an electrophilic species, resulting in heterolytic splitting of the double bond to produce a carbenium ion [40]. The cationic polymerization undergo cationic intermediate as shown below (Equation (16)):

$$\sim\!\sim CH{=}C(H)(OR) \longrightarrow \sim\!\sim CH{-}C^{\oplus}(H)(\ddot{O}R) \longleftrightarrow \sim\!\sim CH{-}C(H)({}^{\oplus}\ddot{O}{=}R) \quad \text{-------------} \rightarrow (16)$$

7.8 Lewis Acids/Friedel – Crafts Catalysts

Lewis acids are the most common compounds used for initiation of cataionic polymerization. The more popular Lewis acids are $SnCl_4$, $AlCl_3$, BF_3 and $TiCl_4$. Although these Lewis acids alone are able to induce polymerization, the reaction occurs much faster with a suitable cation source. The cation source can be water, alcohols, or even a carbocation donor such as an ester or an anhydride. In these systems, the Lewis acids referred to as a co-initiator, an intermediate complex forms which then goes on to react with the monomer unit.

The terminology of proton donor or carbocation donor and Lewis acid refer as initiator and co-initiator respectively [41]. Much of the literature before 1990 used as reverse terminology. Proton donor or carbocation referred as the initiator due to its supplies the proton or cation that ultimately adds to monomer to initiate polymerization. Initiator and co-initiator represent an initiating the polymerization during the reaction to form an initiator-co-initiator complex. It proceeds to donate a proton or carbocation to monomer further to initiate propagation [42, 43].

Initiation by Lewis acids requires and/or proceeds faster in the presence of either a proton donor such as water, hydrogen halide, alcohol and carboxylic acid or carbocation donor such as an alkyl halide such as t-butyl chloride and triphenylmethyl chloride, ester, ether, or anhydride. Polymerization occurs immediately with traces of water in the case of dry isobutylene unaffected in presence of dry boron trifluoride [44].

Lewis acidity for different metals generally increases with increasing atomic number in each group as Ti > Al > B; Sn > Si; Sb > As [45]. For any metal, Lewis acidity increases with increasing oxidation state, for example, TiCl4 > TiCl2. Ligands increase Lewis acidity in the order: F > Cl > Br > I > RO > RCOO > R, Ar (R = alkyl and Ar = Aryl). Strongest Lewis acid such as SbF_5 is not much useful due to fast and uncontrolled polymerization. The activity of complex of initiator – co-initiator increases with increasing acidity of the initiator. The activity of the initiator increases as hydrogen chloride > acetic acid > nitroethane > phenol > water > methanol > acetone in the polymerization of isobutylene with tin (IV) chloride [46, 47].

Carbenium ion additions to isobutene as key steps in the cationic polymerization of isobutene. The dashed arrow corresponds to the overall reaction (Equations (17)–(23)).

$$H_2SO_4 \rightleftharpoons H^+ + HSO_4^- \dashrightarrow \quad (17)$$

$$BF_3 + H_2O \rightleftharpoons H^+\ BF_3(OH^-) \dashrightarrow \quad (18)$$

$$H^{\oplus}(BF_3OH)^{\ominus} + CH_2{=}C(CH_3)_2 \longrightarrow CH_3{-}C^{\oplus}(CH_3)_2\ (BF_3OH)^{\ominus} \dashrightarrow \quad (19)$$

$$CH_2{=}C(CH_3){-}CH_3 \xrightarrow{H^+} CH_2{-}C^{\oplus}(CH_3){-}CH_3 \dashrightarrow \quad (20)$$

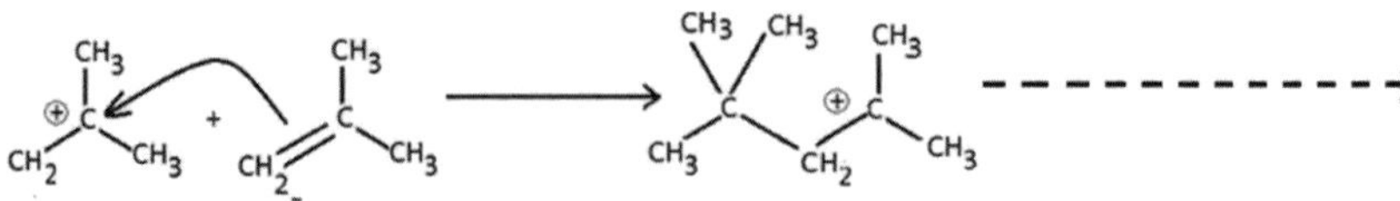

(21)

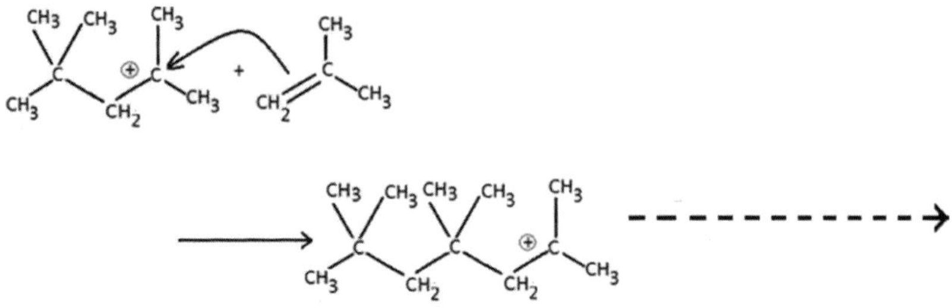

(22)

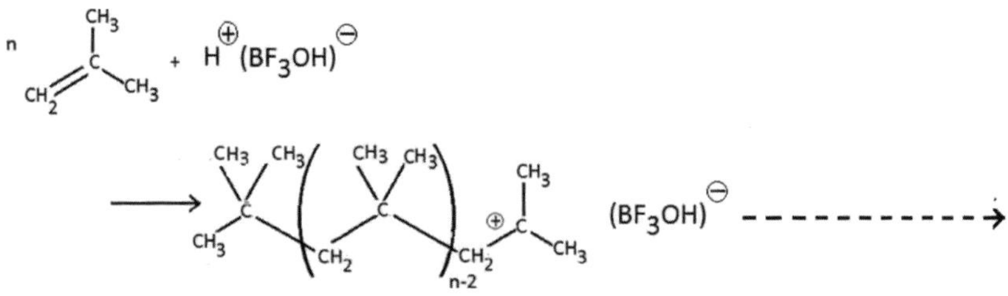

n + $H^{\oplus}(BF_3OH)^{\ominus}$ → $(BF_3OH)^{\ominus}$ (23)

The counter ion produced by the initiator – co-initiator complex is less nucleophilic than that of the Brønsted acid A^- counter ion. Halogens, such as chlorine and bromine, can also initiate cationic polymerization upon addition of the more active Lewis acids. Initiation with boron trifluroide as co-initiator and water as initiator use in Lewis acids catalyzed cationic polymerization.

The order of activity of a series of initiators or co-initiators may differ depending on the identity of the other component, monomer, solvent, or the presence of competing reactions. The activity of boron halides in isobutylene polymerization, BF3 > BCl3 > BBr3, with water as the initiator is the opposite of their acidities. Hydrolysis of the boron halides to inactive products, increasing in the order BBr3 > BCl3 > BF3, is responsible for results [48, 49]. Many of the reactions that do not terminate the growth of a propagating chain. However, terminate the kinetic chain due to generation of new propagating species in the process. Various reactions lead to termination of chain growth in cationic polymerization [50].

7.9 Carbonium Ion Salts

Stable carbonium ions (Equation (24)) used to initiate chain growth of only the most reactive olefins. Well-defined structures produce from the stable carbonium ion salts. The initiator used in kinetic studies due to the case of being able to measure the disappearance of the carbonium ion absorbance. Common carbonium ions are trityl and tropylium cations.

$$(C_6H_5)_3C^{\oplus}A^{\ominus} + CH_2{=}CH(C_6H_5) \rightarrow (C_6H_5)_3C{-}CH_2{-}\overset{\oplus}{C}H(C_6H_5)\ A^{\ominus} \quad (24)$$

There are variety of initiators among them some of the initiators require a co-initiator to generate the needed cationic species. The cationic polymerization is a kind of repetitive alkylation reaction. To initiate cationic polymerization, strong acidic catalysts such as H_2SO_4, $HClO_4$, and H_3PO_4 or Lewis acids such as BF_3, $AlCl_3$, $TiCl_4$, and $SnCl_4$ are used. Lewis acids are by far the most important initiators for industrial cationic polymerizations. Initiation with Lewis acids requires the presence of a trace of proton donor such as water, alcohol, and organic acid or a cation donor such as alkyl halide. BF3 is particularly effective catalyst, requires a trace of water or methanol as a co – catalyst. However, intermediate cations must be stable.

Propagation is usually very fast. However, it runs at low temperatures. The monomer needs substituent that is electron donating in cationic polymerization. It can also have prapagating oxonium ions (Equations (25)–(27)).

$$H_2C{=}CH{-}\ddot{O}R \longrightarrow H_2C{-}\overset{\oplus}{C}H{-}\ddot{O}R \quad (25)$$

$$H_2C{-}\overset{\oplus}{C}H{-}\ddot{O}R + H_2C{=}CH{-}\ddot{O}R \longrightarrow H_2C{-}CH(\ddot{O}R){-}CH_2{-}\overset{\oplus}{C}H{-}\ddot{O}R \quad (26)$$

Carbenium ion or carbocation

$$H_2C{-}CH(:\ddot{O}R){-}CH_2{-}\overset{\oplus}{C}H{-}\ddot{O}R + H_2C{=}CH{-}\ddot{O}R \longrightarrow H_2C{-}CH(:\ddot{O}R){-}H_2C{-}CH(:\ddot{O}R){-}CH_2{-}\overset{\oplus}{C}H{-}\ddot{O}R \dashrightarrow \quad (27)$$

Propagating cation stabilize
by electron
donating nature of oxygen

At high pressure, the ionic intermediates stabilize and addition without elimination may occur. The polymerization uses an initiator and surfactant. The active centers result with further reaction until termination/transfer reactions by water [51, 52]. The initiators of a cationic polymerization of styrene can be strong acids like protonic acids such as perchloric, hydrochloric, or sulfuric acid or Lewis acids such as BF_3. BCl_3, and $AlCl_3$ [53–57]. To initiate the polymerization, additionally alumina, silica and molecular silica used. Depending on the solvent, the degree of ion separation decreased under the influence of electric field [58, 59]. The effect is small in toluene due to the low ionization value of the solvent and intermediate values ionization value, the epichlorohydrin, the highest changes in rate observed. In nitrobenzene, a solvent with high ionization value, the ion separation is complete. Application of an external field does not affect the free ion concentration and the rate. The gradual change in rate of polymerization indicates a radical process, suggests an unimolecular termination characteristic for the cationic mechanism [60–64].

7.10 Polymerization of Isobutylene

The most common cationic polymerization is the polymerization of isobutylene. It polymerizes with Friedel – Crafts catalysts in a reaction. It involves tertiary carbocation intermediates. The reaction is sensitive to temperature, solvent nucleophile impurities. Friedel-Crafts reactions during growth with aromatic solvents significantly decrease molecular weight [65]. Monomer addition increases molecular weight.

The polymerization of styrene initiated by $HClO_4$ [66] involves following steps (Equations (27)–(29)) explained by intermediate perchlorate esters [67]:

1. Fast polymerization shows electrical conductivity and all orange red color due to presence of electrons

2. Characterized by the absence of conductivity and color
3. Conductivity and reappearance of color

Initiation

$$H^+ClO_4^- + CH_2{=}CH(C_6H_5) \longrightarrow CH_3{-}\overset{+}{C}H(C_6H_5)\,ClO_4^- \quad (27)$$

Propagation

$$CH_3{-}\overset{+}{C}H(C_6H_5)ClO_4^- + CH_2{=}CH(C_6H_5) \longrightarrow CH_3{-}CH(C_6H_5){-}CH_2{-}\overset{+}{C}H(C_6H_5)ClO_4^- \quad (28)$$

Termination by unimolecular rearrangement

$$\sim\!\!\sim CH_2{-}\overset{+}{C}H(C_6H_5)ClO_4^- \longrightarrow \sim\!\!\sim CH_2{=}CH(C_6H_5) + H^+ClO_4^- \quad (29)$$

7.11 Termination

Termination generally occurs through unimolecular rearrangement with the counterion (Equation (30)). In the process, an anionic fragment of the counter ion combines with the propagating chain end. This is not only inactivates the growing chain, but also terminates the kinetic chain by reducing the concentration of the initiator – co-initiator complex.

$$H{-}(CH_2{-}CH_2)_n{-}CH_2{-}\overset{\oplus}{C}(CH_3)_2\ \overset{\ominus}{A} \longrightarrow H{-}(CH_2{-}CH_2)_n{-}CH_2{-}C(CH_3)_2{-}A \quad (30)$$

Termination by combination with an anionic fragment from the counterion (Equation (31)).

Termination by chain transfer reaction

$$\sim\!\!\sim CH_2-\overset{+}{C}H(C_6H_5)\,ClO_4^- + CH_2=CH(C_6H_5) \dashrightarrow \quad (31)$$

$$\longrightarrow \sim\!\!\sim CH_2=CH(C_6H_5) + CH_3-\overset{+}{C}H(C_6H_5)\,ClO_4^-$$

(n-Bu)4$N^+ClO_4^-$ is accelerate the polymerization rate in comparison with the rate for salt free polymerization [68]. Friedel-Crafts reactions during growth with aromatic solvents significantly decrease molecular weight [69]. Monomer addition increases molecular weight.

Cationic polymerization is a type of chain growth polymerization, in which a cationic initiator transfers charge to a monomer, which becomes reactive. The reactive monomer undergoes similarly with other monomer to form a polymer. The types of monomers necessary for cationic polymerizations are limited to olefins with electron donating substituents and heterocycles. Similar to anionic polymerization reactions, cationic polymerization reactions are very sensitive to the type of solvent used. Specifically, the ability of a solvent to form free ions will dictate the reactivity of the propagating cationic chain. Monomers for cationic polymerization are nucleophilic and for a stable cation upon polymerization.

Cationic polymerization of olefin monomers occurs with olefins that contain electron-donating substituents. These electron-donating groups make the olefin nucleophile enough to attack electrophilic initiators or growing polymer chains. At the same time, these electron-donating groups attached to the monomer must be able to stabilize the resulting cationic charge for further polymerization.

7.12 Ionic Chain Carriers

Chain growth takes place through the repeated addition of a monomer in a head – to – tail manner to the carbonium ion, with a retention of throughout ionic character. The propagation mechanism depends on the counter ion,

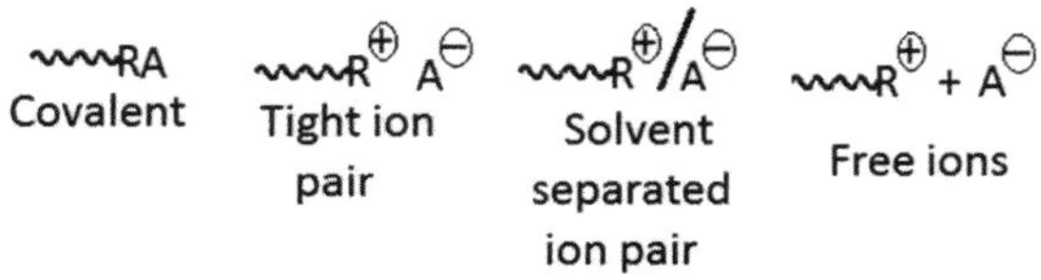

~~~RA — Covalent; ~~~$R^{\oplus}$ $A^{\ominus}$ — Tight ion pair; ~~~$R^{\oplus}$/$A^{\ominus}$ — Solvent separated ion pair; ~~~$R^{\oplus}$ + $A^{\ominus}$ — Free ions

Figure 7.2 Association of ions in solution.

solvent, temperature, and the type of monomer. The solvent and the counter ion have a significant effect on the rate of propagation. The counter ion and the carbonium ion can have different associations (Figure 7.2), ranging from a covalent bond, right ion pair not separated, solvent – separated ion pair i.e. partially separated and free ions i.e. completely dissociated.

The association is strongest as a covalent bond and weakest when the pair exists as free ions. In cationic polymerization, the ions tend to be in equilibrium between an ion pair either tight or solvent separated, and free ions. The more polar the solvent used in the reaction, the better the solvation and separation of the ions. Since free ions are more reactive than ion pairs. Therefore, the rate of propagation is faster in more polar solvents.

The size of the counter ion is also a factor. A smaller counter ion, with a high charge density, will have stronger electrostatic interactions with the carbonium ion than will have a larger counter ion that has a lower charge density. Further, a smaller counter ion is more easily solvated by a polar solvent than a counter ion with low charge density. The result is increased propagation rate with increase solvating capability of the solvent. Chain length affected by temperature. Low reaction temperatures are preferred for producing longer chains.

The ionic polymerization results stereospecific polymerization in which bonds are broken and made at a single asymmetric atom (usually but not necessarily carbon), and which lead largely to a single stereoisomer. However, in solvents with poor solvating power, there may be extensive coordination between initiator, propagating chain end, and monomer, which results in isotactic (or syndiotactic) placements almost exclusively.

Side reactions of cationic polymerization

Cationic polymerization plagued by numerous side reactions. Many of the side reactions (Equations (32)–(36)) lead to chain transfer. It is difficult to achieve high molecular weight due to chain transfer of each initiator can give
~~~

rise to many separate chains. The side reactions can be minimized, however, not possible to eliminate by running the reaction at low temperature.

7.13 Chain Transfer

Chain transfer can take place in two ways, one by hydrogen abstraction from the active chain end to the counter ion. In the process, the growing chain terminated, but the initiator – co-initiator complex regenerates to initiate more chains.

$$\sim\sim\sim CH_2-C^{\oplus}(CH_3)_2 + H_2C=C(CH_3)_2 \longrightarrow \sim\sim\sim CH=C(CH_3)_2 + H_3C-C^{\oplus}(CH_3)_2 \quad (32)$$

$$H\!-\!(CH_2-CH_2)_n\!-CH_2-C^{\oplus}(CH_3)-CH_2-H \;\; A^{\ominus} \longrightarrow H\!-\!(CH_2-CH_2)_n\!-CH_2-C(CH_3)=CH_2 + HA \quad (33)$$

Chain transfer by hydrogen abstraction to the counter ion

The second method involves hydrogen abstraction from the active chain end to the monomer. This terminates the growing chain and also forms a new active carbonium ion – counter ion complex which can continue to propagate, thus keeping the kinetic chain intact.

$$H\!-\!(CH_2-CH_2)_n\!-CH_2-C^{\oplus}(CH_3)-CH_2-H + CH_2=C(CH_3)-CH_3 \longrightarrow H\!-\!(CH_2-CH_2)_n\!-CH_2-C(CH_3)=CH_2 + CH_3-C^{\oplus}(CH_3)-CH_3 \quad (34)$$

Chain transfer by hydrogen abstraction to the monomer.

In analogy to controlled/"living" cationic polymerizations, radical polymerizations can become controlled under conditions in which a low, stationary concentration of the active species is maintained and a fast, dynamic equilibrium is established between the active and dormant species [70–72].

Release of HX

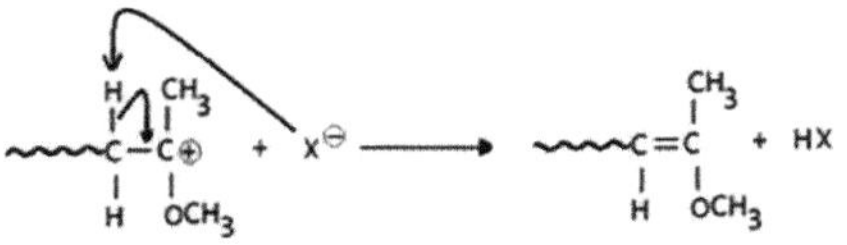

------------→ (35)

Backbiting

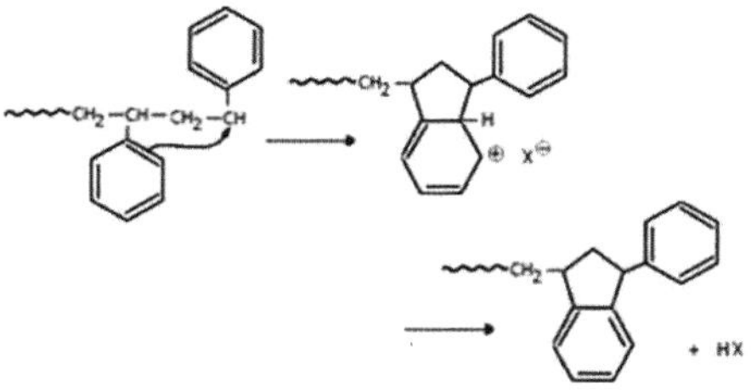

------------→ (36)

7.13.1 Advantages

- Polymer structures defines and proceeds in a controlled manner.
- Technique of synthesizing of well-defined homopolymers as well as different macromolecular architectures, using monomers such as isobutylene, styrene and derivatives, vinyl ethers, etc.

7.13.2 Disadvantages

- The chains have penetrated inside the particles, and cannot undergo further propagation, since the catalyst lock at the interface in many cases.
- Rare case of termination reaction via water is reversible

7.14 Anionic Polymerization

Anionic polymerization is carried out through a carbanion active species. Important way of generating carbon nucleophiles involves removal of proton from a carbon by a base. Anionic polymerization involves monomers that have electron withdrawing groups such as acrylonitrile, vinylchloride, methyl methacrylate, styrene, etc. The monomers polymerized have anionic polymerization if the sites derived which are capable to induce chain growth. Anionic polymerization used to synthesize polymers with narrow molecular weight distributions, with functional groups at the chain ends, etc.

The anions produced are carbanions. The negative charge gives good nucleophilic properties and it used in the formation of new carbon – carbon bond. These ions stabilized by resonance or by inductive effects. Monomers polymerized anion reaction if the sites derived therefrom are capable of induce chain growth. Anionic polymerization of polar monomers – type of polar monomers, potential problems due to side polar groups, stereo regulation in polymerization in methyl methacrylate polymerization. The syntheses rely on the presence of a reactive carbanion at the end of each polymer molecule throughout the reaction, living polymerization. At low pressures, cationic and radical cationic polymerization may proceed through elimination reactions. At high pressure, the ionic intermediates stabilize and addition without elimination may occur.

The propagating chain is a carbanion in anionic polymerization (Equation (37)). Initiator undergoes nucleophilic addition to monomer forms it. Monomers having substituent groups such as nitro, cyano, carboxyl, vinyl and phenyl capable of stabilizing a carbanion through resonance or induction effect. The resonance or induction effect is most susceptible to anionic polymerization.

$$Nu^{\ominus} + H_2C{=}CHY \longrightarrow Nu{-}CH_2{-}\underset{\displaystyle Y}{\underset{|}{CH}} \qquad \text{- - - - - - - - - - - -}\rightarrow \quad (37)$$

Carbon atom of carbanions contains eight electrons in the valence shell. Six electrons from the covalent bond and the other two electrons remain as lone pair with the negative charge. It is a strong nucleophile (Lewis base) and electron rich species (Figure 7.3).

Anionic polymerization is the reaction using the base or nucleophile as an initiator such as $NaNH_2$, $LiN(C_2H_5)$, alkoxides, hydroxides, cyanides, phosphines, amines and organometallic compounds such as n – C_4H_9 Li and C_6H_5– MgBr. Alky lithium is the most useful initiator and used to initiate 1,3 – butadiene and isoprene. It is soluble in hydrocarbon solvents. Initiation proceeds by addition of the metal alkyl to monomer as

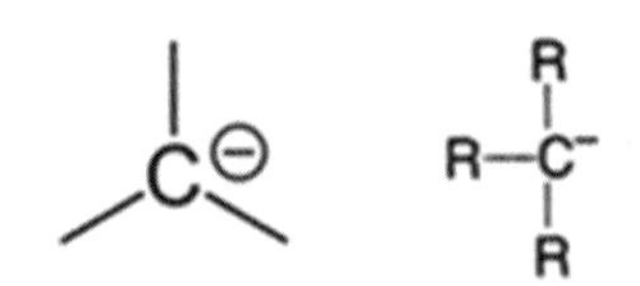

Figure 7.3 Election rich species.

Monomer reactivity increases with increasing ability to stabilize the carbanion charge. Strong nucleophiles such as amide or alkyl carbanion are needed to polymerize monomers such as styrene, 1,3 butadiene with relatively weak electron withdrawing substituents. Weaker nucleophiles, such as alkoxides, and hydroxide ions, can polymerize monomers with strongly electron withdrawing substituents such acrylonitrile, methyl methacrylate, and methyl vinyl ketone.

Anionic polymerizations show many of the same characteristics as cationic polymerizations. However, since the nature of carbanions is different from carbenium ions, there are distinct differences. In contrast to cationic polymerization, neither termination nor chain transfer occur in many anionic polymerizations (living polymerization) especially when polar substances are absent. Anionic active centers are usually much more stable than cationic active centers. Although anionic polymerizations proceed rapidly at low temperatures, they are not usually as temperature sensitive as cationic polymerizations, and polymerizations usually proceed well at ambient temperature and higher. Variety of basic initiators use to initiate anionic polymerizations [73, 74]. Propagation occurs by the successive insertion of monomers into the partial bond between the propagating anion and its cationic counterion.

Anionic polymerization with carefully purified reagents may lead to systems in which chain termination is absent. The polymers arise from such reactions are referred to as living polymers [75, 76]. In practice, various polymers with well-defined structures synthesized by means of this living polymerization.

The anionic polymerization is realized by inserting monomers into the space between cationic and anionic ion-pairs and most important methods for synthesizing polymers with a very narrow molecular weight distribution and specific structures [77–80]. However, the ion-pairs can appear in aggregated ion-pairs, contact ion-pairs, solvent-separated ion-pairs and free ions and there exists a chemical equilibrium between them, which largely shifts toward the former, aggregated ion-pairs. However, the propagation occurs only through the latter [81, 82].

7.15 Ideal Anionic Polymerization

Negative centers repel one another, therefore, termination and recombination is not possible. An ideal polymerization is living in nature that does not terminate until the addition of terminator. Initiation is normally very fast relative to propagation and all chains simultaneously grow. This will results in low polydispersity or monodispersity. The rate of polymerization for many monomer is high even at optimum temperatures. This is partly for the high concentration of the anion centers.

In anionic polymerization, the efficient generation of a significant equilibrium concentration of a carbonion requires proper choice of initiator. The equilibrium will favor carbanion formation only when the acidity to produce anion is to be greater than that of the conjugate acid corresponding to the based used for deprotonation. A group attached to a carbon leaves without its electron pair. Negative ion adds to a carbon – carbon double bond or triple bond. Formation of carbanion as follows (Equations (38)–(40)):

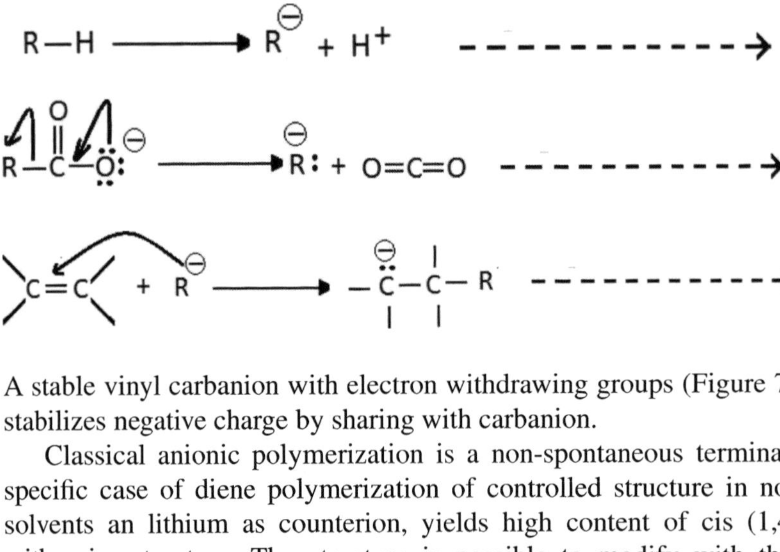

$$R{-}H \longrightarrow R^{\ominus} + H^{+} \tag{38}$$

$$R{-}C({=}O){-}\ddot{O}:^{\ominus} \longrightarrow R:^{\ominus} + O{=}C{=}O \tag{39}$$

$$>C{=}C< + R^{\ominus} \longrightarrow -\overset{\ominus}{\ddot{C}}{-}C{-}R \tag{40}$$

A stable vinyl carbanion with electron withdrawing groups (Figure 7.4) that stabilizes negative charge by sharing with carbanion.

Classical anionic polymerization is a non-spontaneous termination. In specific case of diene polymerization of controlled structure in non-polar solvents an lithium as counterion, yields high content of cis (1,4) units with microstructure. The structure is possible to modify with the introduction of polar additives. The reactions with low propagating rates with increased probability of deactivation as compared to polar solvents. This type non-spontaneous terminations are limited to monomers such as diene, styrene.

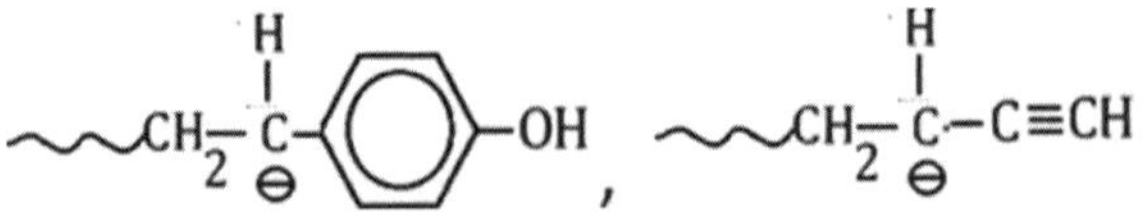

$$\sim\!\sim CH_2{-}\overset{H}{\underset{\ominus}{C}}{-}C_6H_4{-}OH, \quad \sim\!\sim CH_2{-}\overset{H}{\underset{\ominus}{C}}{-}C{\equiv}CH$$

Figure 7.4 Carbanions with electron withdrawing groups.

7.16 Reactivity of Initiators

The reactivity of an initiator in anionic polymerization depends on the

- Nucleophilic nature of the anion

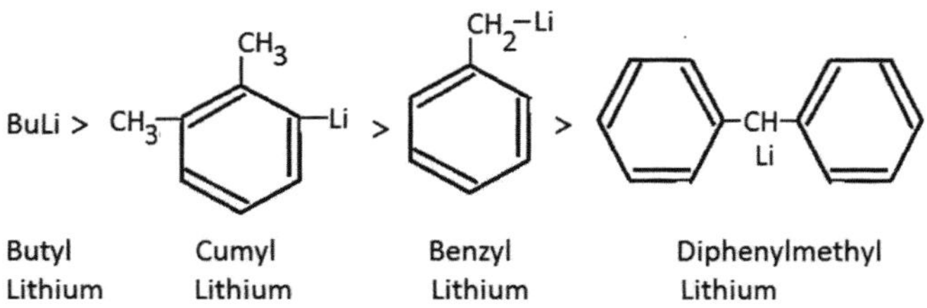

BuLi > Cumyl Lithium > Benzyl Lithium > Diphenylmethyl Lithium

Butyl Lithium | Cumyl Lithium | Benzyl Lithium | Diphenylmethyl Lithium

- Ionic radius of the counterion

$$NR_4^+ > Cs^+ > K^+ > Na^+ > Li^+$$

- Polarity of the solvent

Anionic polymerization of monomers should have the sites derived therefrom they are capable to induce chain growth (Equations (41)–(43)).

- Non-polar vinyl compounds with strong delocalization such as styrene, α – methyl styrene, vinyl naphthalene, butadiene, isoprene, cyclohexadiene, etc.

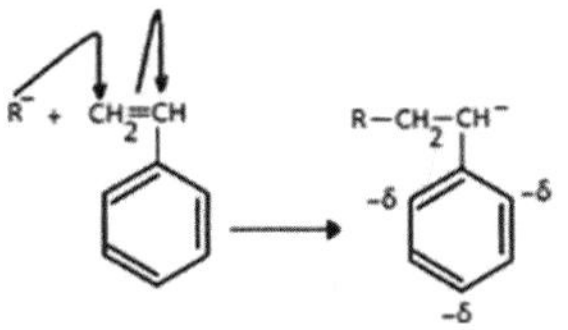

$R^- + CH_2{=}CH(C_6H_5) \longrightarrow R{-}CH_2{-}CH^-(C_6H_5)$ (41)

- Polar compounds with electron attracting substituents such as vinyl pyridine, acrylates, vinyl ketones, acrolein, etc.

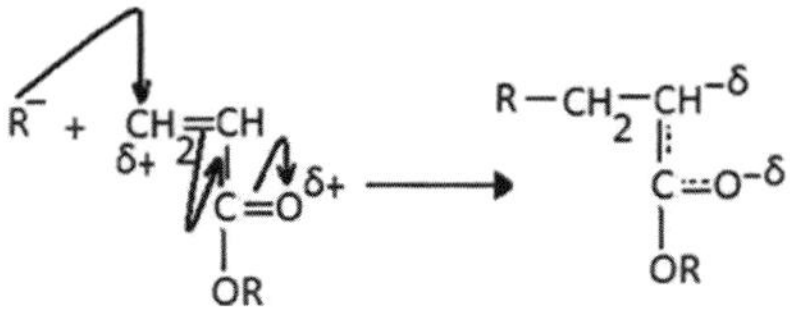

$R^- + CH_2{=}CH{-}C(=O)OR \longrightarrow R{-}CH_2{-}CH^{-\delta}{-}C{=}O^{-\delta}(OR)$ (42)

- Isocyanates, R – N = C =O. isocyanides, etc.

$$-N{=}C{=}O + R^{\ominus} \longrightarrow -N(H)-C(=O)-R \quad (43)$$

7.17 Reaction Process of Polystyrene

Styrene monomer contains phenyl group to act as an electron withdrawing or an electron-donating center. The growth end of the polymer may be either a carbenium ion or a carbanion. Anions from styrene monomer form during reaction with organo lithium compounds as initiators [83, 84]. The initiation is much faster than propagation with instantaneously anions formation (Equation (44)).

$$R{-}Li + H_2C{=}CH(C_6H_5) \longrightarrow R{-}H_2C{-}\overset{\ominus}{C}H(C_6H_5)\ Li^{\oplus} \quad (44)$$

1. The reaction of 1,3, bis(1-phenylvinyl)benzene with organolithium and styrene yields a dianion with good solubility in hydrocarbon solvents [85–87]. Addition of bifunctional initiator is formed by reaction of 1,3, bis(1-phenylvinyl)benzene with organolithium, yielding a dianion with good solubility in hydrocarbon [88, 89]. A similar system mentioned below [90]. The propagation reaction occurs as follows (Equation (45)):

Propagation

$$Bu{-}\overset{\ominus}{C}H(C_6H_5)\ Li^{\oplus} + (n-1)\ CH_2{=}CH(C_6H_5) \longrightarrow Bu{-}\left(CH_2{-}CH(C_6H_5)\right)_{n-1}{-}CH_2{-}\overset{\ominus}{C}H(C_6H_5)\ Li^{\oplus} \quad (45)$$

Where Bu = Butyl

7.18 Reaction Process with Alkene with Functional Groups

Mere alkene will not undergo anionic polymerization due non-presence of electron withdrawing groups [91–94]. Potassium amide as an initiator with an electron with drawing molecule, in the initiation step, the base adds to double bond to form a carbanion (Equation (46)).

$$K^+NH_2^- + CH_2{=}CH(X) \longrightarrow H_2N{-}CH_2{-}\bar{C}H(X)\ K^+ \quad (46)$$

Where X = electron withdrawing group

In chain propagation, the carbanion adds to the double bond and the process repeats to form a polymeric carbanion (Equation (47)].

$$H_2N{-}CH_2{-}\bar{C}H(X)\ K^+ + nCH_2{=}CH(X) \longrightarrow H_2N{-}CH_2{-}\left[CH_2{-}CH(X)\right]_n{-}\bar{C}H(X)\ K^+ \quad (47)$$

The chain reaction terminated by addition of an acid (Equation (48)).

$$H_2N{-}CH_2{-}\left[CH_2{-}CH(X)\right]_n{-}\bar{C}H(X)\ K^+ \xrightarrow{H^+} H_2N{-}CH_2{-}\left[CH_2{-}CH(X)\right]_n{-}CH_2(X) \quad (48)$$

Polymers such as polystyrene, poly(acrylonitrile), poly(ethylene oxide), poly(methylmethacrylate) etc. undergo anionic polymerization.

7.18.1 Advantages

- The polymer anions use to prepare for the system of diblock, triblock and multiblock copolymers [95–97].
- Livingnature of the chain is that it converts into interesting functional end groups [98–101].
- Anionic polymerization are more reproducible due to the reaction components are better defined and more easily purified.
- Anionic polymerization offers potentially better control of stereochemistry of polymers.

7.18.2 Disadvantages

- Limited monomers to be polymerized anionically.
- Require pure monomer.

- Polymerizations must be conducted under inert conditions, i.e., in the absence of moisture or other electrophiles.
- Oxygen undergoes electron transfer with carbanions, leading to chain coupling [102]. Airtight experimental condition is required to avoid the entry to oxygen.

7.19 Ring Opening Polymerization

Ring opening polymerization reactions constitute an important class of polymerization. These reactions utilized for preparation of polyamides, aliphatic polyesters, etc. Strained monomeric rings, medium strained rings, and oligomeric large ring systems have high, medium or little/no reaction exotherms respectively.

Ring opening polymerization can occur by step-growth or chain growth. Chain growth process involves repeated addition of monomer to the chain end leads to an increase in the molecular weight. Cyclic monomers constitute a broad family of monomers. It is able to polymerize by anionic or related nucleophilic polymerization techniques. The polymerization of cyclic ether, esters, amide and cyclic monomers such cycloalkanes, cyclic amines, utilizes step-growth or chain growth. Chain growth process involves repeated addition of monomer to the chain end leads to an increase in the molecular weight. Commercially important ring opening polymerization has three factors that will be:

- Nature of the heteroatom, its electronegativity and bond strength with the carbon atom
- Size of the ring
- Steric factors.

The order of the basics of cyclic ethers is as follows:

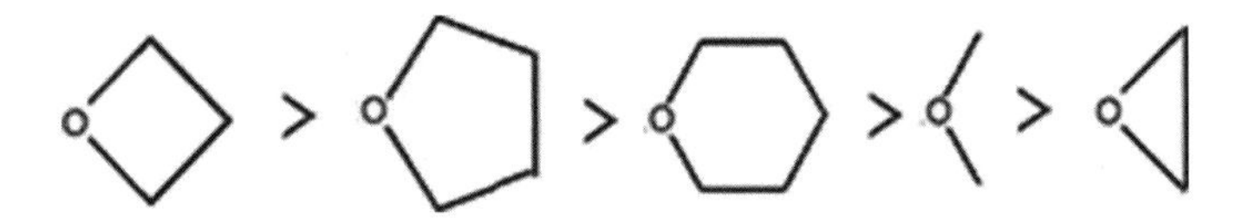

Ring size of the cyclic monomer has a controlling effect both on the preparation of the monomer, and on its stability and polymerizability. All ring opening polymerization virtually involve ring chain equilibrium from measurable quantities of monomeric or oligomeric cyclics. Medium sized rings of 5–9 members have less ring strain. Therefore, lower reaction exotherms during polymerization. Ring opening polymerization undergoes initiation as ring cleavage, propagation as attachment of cyclic monomers and termination ends with polymer (Equations (49) and (50)).

$$R\ddot{O}^{-} + \overset{O}{\triangle}{-}CH_3 \longrightarrow RO{-}CH_2{-}\underset{\displaystyle CH_3}{\underset{|}{CHO^{-}}} \quad \text{------------}\rightarrow (49)$$

$$RO{-}CH_2{-}\underset{\displaystyle CH_3}{\underset{|}{CHO^{-}}} + \overset{O}{\triangle}{-}CH_3 \longrightarrow RO{-}CH_2{-}\underset{\displaystyle CH_3}{\underset{|}{CHO}}{-}CH_2{-}\underset{\displaystyle CH_3}{\underset{|}{CHO^{-}}} \quad \text{------------}\rightarrow (50)$$

Thermodynamics of ring-opening polymerization drive by the enthalpy of the ring opening. The kinetics and selectivity of the ring opening process strongly influences by the nature of the reactive chain ends, the monomers, and the presence of catalysts. Catalytic steps in a ring opening polymerization must occur with the correct relative rates to yield a well-controlled reaction is impressive (Equation (51)). It is possible under this condition that rates of initiation and propagation are higher than termination and inter and intra chain reactions exquisite control over the molecular weight and is distribution is possible [103].

$$\text{(O, O, } H_2C, CH_3, CH_3 \text{ lactone)} \xrightarrow{\text{cat.}} \left[O{-}CH_2{-}\underset{\displaystyle CH_3}{\overset{\displaystyle CH_3}{C}}{-}\overset{\displaystyle O}{\overset{\|}{C}} \right]_n \quad \text{------------}\rightarrow (51)$$

The second class of addition polymerization reactions is one wherein a ring compound breaks and the individual units join linearly. Ring compound breaks and the individual units join linearly. Ring opening polymerization of caprolactam as below:

$$\underset{\text{ε-caprolactum}}{\text{(O, N, H ring)}} \longrightarrow \underset{\text{Nylon 6}}{HO\left[\overset{O}{\overset{\|}{C}}\sim\sim NH\right]_n H} \quad \text{------------}\rightarrow (52)$$

7.19.1 Advantages

- Ring opening polymerization is a simple technique. However, it requires thermal insulation.
- The polymer obtains in pure form.
- Using chain transfer agents, it changes directly molecular weight distribution.

7.19.2 Disadvantages

- Highly exothermic and difficult to control heat. Mixing the ingredients is difficult due the viscosity of reaction mass increases.
- Broad molecular weight distribution due to high viscosity and lack of heat transfer.

7.20 Coordination Polymerization

Coordination polymerization is a chain growth polymerization. Coordination polymerization started with the advent of transition metal catalysts, number of polymer reactions utilizes in production of polymers such as polyethylene, polypropylene, etc. Many industrial processes require either homogeneous or heterogeneous catalysts. The active species are coordination complex.

7.21 Ziegler–Natta Polymerization

Recent development in transition metal complexes can function as catalysts for the stereo-selective Ziegler–Natta polymerization of propylene has provided new polypropylene through extensive manipulations about the metal center of the ligand environment [104–107]. The polymerization of vinyl monomers under mild conditions using Lewis acid catalysts to give a stereo regulated, or tactic, polymer [108–116]. Ziegler–Natta catalysts are special type with catalyst with the co-catalyst, which are organometallic compound. Ziegler–Natta catalysts based on the compound $TiCl_4$ or $TiCl_3$ in combination with aluminum chloride as $TiCl_4/R_3Al$, $TiCl_3/R_2AlCl$, and as Grignard regent as $TiCl_4/R_3Al/MgCl_2$.

Trialkyl aluminum acts as the electron acceptor whereas the electron donor is titanium halides. Therefore, it readily combines and forms coordination complexes.

$$R_2Al(\mu\text{-}R)(\mu\text{-}Cl)TiCl_2 + CH_2{=}CHCH_3 \longrightarrow Al(\mu\text{-}R)(\mu\text{-}Cl)Ti \leftarrow CH_2{=}CHCH_3 \quad (53)$$

The complex formed acts as the active center. The monomer attaches with the metal ion of the active center. It forms a π complex with titanium metal. The alkyl group and titanium produces an electron deficient point, attracts the π electrons pair or the monomer, and forms bond. The transition state gives the chain growth at the metal carbon and regenerating the active center.

Titanium halide and an ethyl aluminum compound combine to place an ethyl group on titanium. It gives an active catalyst. Titanium has one or more vacant coordination sites as an empty orbital.

Ethylene reacts with the active form of the Ziegler–Natta catalyst. The π orbital of ethylene overlaps with its two elections with vacant titantium oribital to bind ethylene as a ligand to titanium.

$$CH_3CH_2\text{-}TiCl_n + CH_2{=}CH_2 \longrightarrow CH_3CH_2\text{-}TiCl_n \cdots (CH_2{=}CH_2) \quad (54)$$

Electrons flow from ethylene to titanium increases with electron density at titanium. This weakens the titanium and ethylene bond. Thereby the ethyl group migrates from titanium to one of the carbons of ethylene and become n-butyl titanium halide.

$$CH_3CH_2\text{-}TiCl_n \cdots (CH_2{=}CH_2) \longrightarrow CH_3CH_2\text{-}CH_2\text{-}CH_2\text{-}TiCl_n \quad (55)$$

Continuation of addition of ethylene monomer leads to polyethylene at the end. The active site of the growing chain is the carbon atom directly bonded to the metal.

$$CH_3CH_2\text{-}CH_2\text{-}CH_2\text{-}TiCl_n \xrightarrow{(n\text{-}2)CH_2=CH_2} \text{-}(CH_2\text{-}CH_2)_n\text{-} \quad (56)$$

7.21.1 Advantages

- Polymer structure control achieved in the polymerization of readily available and cheap monomers such as ethylene and propylene.

- Controlling the molecular architecture of polymers, e.g., stereoregularity, is of practical importance both in the development of new polymers or tailor-made polymers and in the control of polymer properties.
- Branching will not occur due to no radical involvement.

7.21.2 Disadvantages

- Ziegler–Natta Catalysts in the polymerization process one catalyst-one material has limitations for tuning of physical properties of polymer through minor adjustments about a given microstructure or accessing completely different microstructure.
- Labor-intensive synthetic process of structural variants that may or may not yield a desired microstructure.

7.22 Summary

- In radical polymerization, the initiator is a radical and the propagation is a carbon radical.
- In cationic polymerization, the initiator is an acid and the propagation is a carbocation.
- In anionic polymerization, the initiator is a nucelophile and the propagation is a carbanion.
- Perfect polymerization would require the consideration of possible reactions.
- To achieve a desired result, polymerization system would able to provide for the monomers that to be added to a solution or emulsion.
- Polymerization methods are unclear until all possibilities are considered.
- Current polymerization system provides an aid for determining the mechanism by which monomers will interact to provide final products.
- Initiation, propagation and termination steps are in chain growth polymerization.
- Radical polymerization applied only to addition polymerization and polyaddition.
- Polycondensations usually proceed with an ionic mechanism through an acid or base catalyzed reaction [117].
- For an alkene to gain electron, strong electron-withdrawing groups such as attachment of nitro, cyano, or carbonyl to the carbons in the double bond is must.

- The chain length formed during the early stages of radical polymerization is high. It will notwithstanding the gel or Trommsdorff effect.
- Reduction of monomers with conversion occurs due to the depletion of monomer.
- Termination made it impossible to get polymers with monodispersed chain lengths and controlled molecular architecture.
- Monomer forms high molecular weight polymer only under extreme conditions.

References

[1] W. Kern, R. C. Schulz, and D. Braun. *J. Polym. Sci.*, **48**, 91 (1960).

[2] P. Ferruti, A. Bettelli, and A. Fer. *Polymer,* **13**, 462 (1972).

[3] H. G. Batz, G. Franzmann, and H. Ringsdorf. *Angew. Chem.*, **84**, 1189 (1972).

[4] H. G. Batz, G. Franzmann, and H. Ringsdorf. *Angew. Chem. Int. Ed. Engl.*, **11**, 1103 (1972).

[5] G. Odian. *Principles of Polymerization*, 4th ed., Wiley-Interscience, New York, NY (2004).

[6] R. Faust, T. D. Shaffer, Eds. *Cationic Polymerization: Fundamentals and Applications, ACS Symposium Series 665*, American Chemical Society, Washington, DC, (1997).

[7] K. Matyjaszewski and C. Pugh. *Cationic Polymerizations; Mechanisms, Synthesis, and Applications*, (Matyjaszewski, K. ed.) Marcel Dekker, New York, NY, (1996).

[8] J. P. Kennedy, and B. Ivan. *Designed Polymers by Carbocationic Macromolecular Engineering: Theory and Practice*, Hanser, New York, NY, (1992).

[9] G. Moad, and D. H. Solomon. *The Chemistry of Radical Polymerization*, 2nd ed. Elsevier, Oxford, 1–9 (2006).

[10] C. H. Bamford, W. G. Barb, A. D. Jenkins, and P. F. Onyon. (1958).

[11] A. Matsumoto, S. J. Nakamura. *Appl. Polym. Sci.*, **74**, 290 (1999).

[12] D. Braun, and E. Manger. *Angew. Makromol. Chem.*, **145**, 101 (1986).

[13] Y. D. Semchikov, A. V. Ryabov, N. L. Khvatova, E. N. Mil'chenko. *Vysokomol. Soedin. Ser. A,* **15**, 451 (1973).

[14] W. Kuran, S. Pasynkiewicz, R. Nadir, and Z. Florjanczyk. *Makromol. Chem.*, **178**, 1881 (1977).

[15] M. J. Kamachi. *Polym. Sci. Polym. Chem. Ed.*, **20**, 1489 (1982).

[16] E. L. Madruga, and J. S. Roman. *MakromolekChem: Rapid Commun.*, **B7**, 307 (1986).
[17] Kh. S. Bagdasar'yan. *Theory of Radical Polymerization*, Davey, Hartford, CT, (1968).
[18] C. H. Bamford, and C. F. H. Tipper. *Comprehensive Chemical Kinetics*, Vol. **14A**, Elsevier, New York, NY, (1976).
[19] G. Moad and D. H. Solomon. *Aust. J. Chem.* **43**, 215 (1990).
[20] K. Matyjaszewski. *Controlled Radical Polymerization; ACS Synposium Series 685*, American Chemical Society, Wasington, DC, (1998).
[21] M. K. Mishra and Y. Yagci. *Handbook of Radical Vinyl Polymerization*, Marcel Dekker, New York, NY, (1998).
[22] G. Allen and J. C. Bevington. "Chain polymerization I," in *Comprehensive Polymer Science*, Vol. **3**, (Eastmond, G. C., Ledwith, A., Russo, S., and Sigwalt, P., eds.) Pergamon Press, New York, NY, (1990).
[23] M. Stickler, D. Panke, and A. E. Hamielec. *J. Polym. Sci., Polym. Chem. Ed.*, **22,** 2243 (1984).
[24] C. H. L. Johnson, G. Moad, D. H. Solomon, T. Spurling, and D. J. Uearing. *J. Aust. J. Chem.*, 43, 1215 (1990).
[25] G. Moad, and E. Rizzardo. *Macromolecules*, **28**, 8722 (1995).
[26] G. A. J. Mortimer. *Polym. Sci. Part A-1* **10**, 163 (1972).
[27] J. Grotewold, and M. M. J. Hirschler. *Polym. Sci. Polym. Chem. Ed.*, **15**, 393 (1977).
[28] T. Corner. *Adv. Polym. Sci.*, **62**, 95 (1984).
[29] S. R. Turner. *Polym. Mater. Sci. Eng.*, **68**, 2 (1993).
[30] D. Colombani. *Chaumont P. Prog. Polym. Sci.*, **21**, 439 (1996).
[31] H. S. Bagdasarian. *Theory of Radical Polymerization*, Izd. Akademii Nauk, Moscow (1959).
[32] K. Matyjaszewski and S. G. Gaynor. *Applied Polymer Science*. (Craver, C. D., Carraher, C. E. Jr eds) Pergamon Press, Oxford, 929, (2000).
[33] D. H. Solomon, and G. Moad. *The Chemistry of Free Radical Polymerization*. Pergamon Press, Oxford (1995).
[34] K. Matyjaszewski, and T. P. Davis. *Handbook of Radical Polymerization*, Wiley-Interscience, Hoboken, NJ (2002).
[35] I. Majoros, J. P. Kennedy, T. Kelen, T. M. Marsalko, *Polym. Bull.*, **31**, 255 (1993).
[36] M. Gyor, H. C. Wang, R. J. Faust, *Macromol. Sci.*, **A29**, 639 (1992).
[37] T. D. Shaffer. US Patent 5, 350, 819 (1994).

[38] M. Szwarc. *Ionic Polymerization Fundamentals*, Hanser, New York, NY (1996).

[39] M. Szwarc, and M. Van Beylen. *Ionic Polymerization and Living Polymers*, Chapman & Hall, New York, NY (1993).

[40] M. Sawamoto and M. Kamigaito. *New Methods of Polymer Synthesis,* (Ebdon, J. R. and Eastmond, G., Ed.) Chapman Hall, London (1995).

[41] G. A. Olah. *J. Am. Chem. Soc.* **94**, 808 (1972).

[42] J. P. Kennedy, and E. Marechal. *Carbocationic Polymerization*, Wiley-Interscience, New York, NY (1982).

[43] D. J. Dunn. "The cationic polymerization of vinyl monomers," in *Chap. 2 in Developments in Polymerization-1*, (Haward, R. N. ed.), Applied Science, London, 1979.

[44] J. P. Kennedy. *J. Polym. Sci. Symp.*, **56**, 1 (1976).

[45] A. G. Evans and G. W. Meadows. *Trans. Faraday Soc.* **46**, 327 (1950).

[46] J. P. Kennedy. *J. Polym. Sci.*, **A-1**, 3139 (1968).

[47] P. H. Plesch. "Isobutene," in *Chap. 4 in The Chemistry of Cationic Polymerizations*, (Plesch, P. H. ed.) Macmillan, New York, NY (1963).

[48] J. P. Kennedy. S. Y. Huang and S. C. Feinberg. *J. Polym. Sci. Polym. Chem. Ed.*, **15**, 2801–2869 (1977).

[49] J. P. Kennedy and S. C. Feinberg, *J. Polym. Sci. Polym. Chem. Ed.*, **16**, 2191 (1978).

[50] A. Gandini, and H. Cheradame, "Cationic Polymerization," in *Encyclopedia of Polymer Science and Engineering*, Vol. 2, (Mark, H. F., Bikales, N. M., Overberger, C. G., and Menges, G., eds) Wiley-Interscience, New York, NY, 729–814, (1985).

[51] S. Cauvin, A. Sadoun R. Dos Santos, J. Belleney, F. Ganachaud, and P. Hemery, *Macromolecules,* **35**, 7919–7927 (2002).

[52] S. Cauvin and R. Dos Santos, F. Ganachaud. *e-Polymers*, 50 (2003).

[53] A. R. Mathieson, A. R. *The Chemistry of Cationic Polymerization*, (Plesch, P. H., ed.), Macmillan, New York, NY 235 (1963).

[54] D. J. Dunn. *Developments in Polymerization*, Vol. **1** (Haward, R. N., ed.), Applied Science Publishers, Barking, 46 (1979).

[55] A. Ledwith and D. C. Sherrington. *Reactivity, Mechanism and Structure in Polymer Chemistry*, (Jenkins, A. D., and Ledwith, A., eds.), Wiley-Interscience, New York, NY 252 (1974).

[56] A. Gandini and H. Cheradame. *Adv. Polym. Sci.*, **35**, 202 (1980).

[57] J. P. Kennedy. *Cationic Polymerization of Olefins: A Critical Inventory*, Wiley, New York, NY 228 (1975).

[58] N. Ise. *Adv. Polym. Sci.*, **6**, 347 (1969).

[59] I. Sakurada, N. Ise and Y. Hayashi. *J. Macromol. Sci. Chem. Part A*, **1**, 1039 (1967).
[60] J. F. Westlake and R. Y. Huang. *J. Polym. Sci. Polym. Chem. Ed.*, **10**, 3053 (1972).
[61] F. Williams. *Fundamental Process in Radiation Chemistry*, (Ausloos, P., ed.), Wiley, New York, NY 515 (1968).
[62] R. C. Potter, C. Schneider, M. Ryska and D. O. Hummel. *Angew. Chem. Intern. Ed.*, **7**, 845 (1968).
[63] F. Williams. *Quart. Rev.*, **17**, 101 (1963).
[64] A. Chapiro and V. Stannet. *J. Phys. Chem.*, **57**, 55 (1960).
[65] M. Kamardi and H. Miyama. *J. Polym. Sci. Polym. Chem. Ed., Part A 1*, **6**, 1537 (1968).
[66] A. Gandini and P. H. Plesch. *Proc. Chem. Soc.*, **3**, 240 (1964).
[67] K. Matyjaszewski. *Macromol. Chem. Macromol. Symp.*, **14**, 433 (1988).
[68] T. Matsuda and T. Higashimura. *J. Polym. Sci. Part A-1*, **9**, 1563 (1971).
[69] M. Kamardi and H. Miyama. *J. Polym. Sci. Polym. Chem. Ed., Part A 1*, **6**, 1537 (1968).
[70] K. Matyjaszewski, Ed. *Cationic Polymerizations: Mechanisms, Synthesis, and Applications*, Marcel Dekker, New York, NY (1996).
[71] K. J. Matyjaszewski. *Phys. Org. Chem.*, **8**, 197 (1995).
[72] D. Greszta, D. Mardare and K. Matyjaszewski. *Macromolecules*, **27**, 638 (1994).
[73] D. H. Richards. "Anionic polymerization," in *Development in Polymerisation-1*, (Haward, R. N ed.) Applied Science Publishers, Essex (1979).
[74] S. Bywater. "Anionic Polymerization," in *Encyclopedia of Polymer Science and Engineering*, Vol. **2**, Wiley-Interscience, New York, NY, 1 (1985).
[75] M. Szwarc, M. Levy, and R. Milkovich. *J. Am. Chem. Soc.* **78**, 2656 (1956).
[76] M. Szwarc. "Living polymers," in *Encyclopedia of Polymer Science and Technology*, Vol. **8**, Wiley-Interscience, New York, NY, 303 (1968).
[77] A. Touris, J. W. Mays and N. Hadjichristidis. *Macromolecules* **44**, 1886 (2011).
[78] A. Touris and N. Hadjichristidis. *Macromolecules* **44**, 1969 (2011).

[79] A. Hirao, S. Tanaka R. Goseki, and T. Ishizone. *Macromolecules* **44**, 4579 (2011).
[80] R. P. Quirk and S.-F. Wang, and M. D. Foster. *Macromolecules* **44**, 7538 (2011).
[81] M. Szwarc, and M. V. Beylen. *Ionic Polymerization and Living Polymers*, Chapman and Hall, New York, NY (1993).
[82] H. L. Hsieh and P. R. Quirk. *Anionic Polymerization*, Marcel Dekker, New York, NY (1996).
[83] H. Hsieh. *J. Polym. Sci. Part A*, **3**, 163.
[84] J. E. L. Roovers and S. Bywater. *Macromolecules*, **8**, 251 (1975).
[85] L. H. Tung, G. Y. S. Lo, J. W. Rakshys and D. E. Beyer. *DOS 2,634,391 to Dow Chemical Comp.; C.A.*, **86**, 190663q (1977).
[86] L. H. Tung, G. Y. S. Lo and D. E. Beyer. *JP 54,063,186 to Dow Chemical Comp.; C.A.*, **91**, 158857y (1979).
[87] L. H. Tung, G. Y. S. Lo, and D. E. Beyer. *Macromolecules*, **11**, 616 (1978).
[88] L. H. Tung, G. Y. S. Lo, J. W. Rakshys and D. E. Beyer. *DOS 2,634,391 to Dow Chemical Comp.; C.A.* **86**, 190663q (1977).
[89] L. H. Tung, G. Y. S. Lo and D. E. Beyer. *Macromolecules* **11**, 616 (1978).
[90] P. Guyot, J. C. Favier, H. Uytterhoeven, M. Fontanille and P. Sigwalt. *Polymer*, **22**, 1724 (1981).
[91] M. Morton and F. R. Ells. *J. Polym. Sci.*, **61**, 25 (1962).
[92] M. Morton and L. J. Fetters. *Macromolecular Reviews*, (Peterlin, A. ed.), Vol. **2**, Interscience, New York, NY 71 (1967).
[93] M. Morton. *Anionic Copolymerization, in Copolymerization* (Ham, G. E. ed.), Interscience, New York, NY 421 (1964).
[94] M. Morton and L. J. Fetters. *J. Polym. Sci., Part A*, **2**, 3311 (1964).
[95] A. Noshay and J. McGrath. *Block Copolymers: Overview and Critical Survey*, Academic Press, Orlando, FL. (1977).
[96] M. J. Folkes. *Processing, Structure and Properties of Block Copolymers*, Applied Science Publishers, Barking (1985).
[97] M. Morton and L. J. Fetters. *Anionic Polymerization: Principles and Practice*, Academic Press, Orlando, FL (1969/1983).
[98] M. Morton and L. J. Fetters. *Macromol. Rev.*, **2**, 71 (1967).
[99] S. Bywater. *Progr. Polym. Sci.*, **4**, 54 (1974).
[100] L. J. Fetters. *J. Polym. Sci. Polym. Symp.*, **26**, 22 (1969).

[101] D. H. Richards, G. C. Eastmond and M. J. Stewart. *Telechelic Polymers: Synthesis and Application*, (Goethals, E. J., ed.), CRC Press, Boca Raton, FL., 33ff (1989).
[102] R. P. Quirk and W.-C. Chen. *J. Polym. Sci., Part A Polym. Chem.*, **22**, 2993 (1984).
[103] M. K. Kiesewetter, E. J. Shin, J. L. Hedrick, and R. M. Waymouth. *Macromolecules*, **43**, 2093–2107 (2010).
[104] H. H. Brintzinger, D. Fischer, R. MEllhaupt, B. Rieger, and R. M. Waymouth. *Angew. Chem.*, **107**, 1255 (1995).
[105] L. Resconi, L. Cavallo, A. Fait and F. Piemontesi. *Chem. Rev.*, **100**, 1253 (2000).
[106] G. MEller, and B. Rieger. *Prog. Polym. Sci.*, **27**, 815 (2002).
[107] C. De Rosa, F. Auriemma, A. Di Capua, L. Resconi, S. Guidotti, I. Camurati, I. E. NifantGev and I. P. Lashevtsev. *J. Am. Chem. Soc.*, **126**, 17040 (2004).
[108] K. Ziegler et al. *Angew. Chem.* **67**, 426–541 (1955).
[109] G. Natta. *Angew. Chem.* **68**, 393 (1956).
[110] a) K. Ziegler. *Angew. Chem.* **71**, 623 (1959); b) K. Ziegler. *Angew. Chem.* **72**, 829 (1960).
[111] C. L. Arcus. *Progress in Stereochemistry,* Vol. **3**, (de la Mare, P. B. D., and Klyne, W. Eds). Butterworth Inc., Washington, DC, 269–288, (1962).
[112] M. N. Berger et al. *Adv. Catalysis* **19**, 211 (1969).
[113] T. Keii. *Kinetics of Ziegler-Natta Polymerization,* Halsted Press, New York, NY, 129–162 (1973).
[114] R. N. Haward, Ed. *Developments in Polymerization*, Vol. **2**, Burgess-Intl., Philadelphia, PA, 81–148 (1979).
[115] H. J. Sinn and W. Kaminsky. *Advan. Organomet. Chem.*, **18**, 207 (1980).
[116] D. M. P. Mingos. *Comp. Organometal. Chem.*, **3**, 72–75 (1982).
[117] a) C. S. Marvel and P. H. Aldrich. *J Am Chem Soc.*, **81**, 1978 (1959); b) C. S. Marvel, L. E. Olson. *J. Polym. Sci.,* **26**, 23 (1957).

8

Polymer Synthesis

8.1 Introduction

Polymer synthesis starts from the elegance of nature with laboratory synthesis. Polymer synthesis is to describe the process of assembling monomers by covalent bonds. The synthesis is not only for the complexity of the targeted polymer, but also for the considerable stereochemical control that accompanied in the polymer. The success of the polymer relates to the introduction of fossil feed stocks as a basis for synthesis. Polymer synthesis starts with great number of monomer as starting material. Not only monomers but also their substituted derivative also undergoes polymerization reactions to improve materials properties [1, 2]. Hence, polymer synthesis constitutes a large proportion of commercially manufactured organic materials.

The evolution of synthetic methods have emerged from

- The deliberate attempt to perfect polymerization reaction or invent a new method in order to permit the preparation of target polymer that may be synthetic product.
- Studies of reactivity of new class of polymerization such as introduction of organometallic compounds or the classical elemental substances from organic sources such as new class of carbon derivatives.

Polymer chemistry has played a central and decisive role in polymer synthesis that becomes most expressive branch in view of its creating new polymers with unlimited scope. Polymer chemistry drives forward and sharpens ability to create polymer through polymer synthesis from which to choose polymer for each application. The most complex and challenging polymerization reactions become the prime driving force for the advancement of the art and science of polymer synthesis [3].

Polymer synthesis with several specific methods does not have general applicability. However, most of methods lead to products that cannot be easily obtained by the more general methods. These methods are valuable tools in polymer synthesis. The current state of the art in polymer synthesis relies on polymerization methods [4].

8.1.1 Requirements

Polymers synthesize follows different mechanisms and techniques. Several problems associate to their synthesis. The size of molecules may also have a dramatic effect on the physical properties [5]. Besides the behaviour of the polymer, due to the functional group presents in the substrate and monomer also show peculiar properties that derive from their own structure. It confers to the polymer properties typically basic properties from possibility of chain with linear or branched or both.

In polymer synthesis, the major requirements are

1. Ease of preparation with controlled degree of polymerization
2. Compatibility with most organic reagents
3. Stability with respect to chemical and mechanical properties.
4. Inertness of the backbone of the polymer towards reactants and regents.
5. Low cost and commercial availability of the monomer.

The major route to synthesis that leads to the polymerization techniques are:

- To connect two or more of same or different molecular entities via monomer
- The possibility of monomer substitution by a functional group
- Introduction of small weight fractions of a comonomer having functionality higher than two [6].

8.1.2 Essentials of Polymer Synthesis

The unique physical and chemical properties of polymers are extremely useful for a wide range of polymer applications. The points of synthesis and research are normally based on

- Synthetic polymers use to be from relative inertness to biodegradation.
- Concerns about preventing or retarding attack on polymers by bacteria, fungi, insects, rodents and other animals.
- To eliminate problems encountered with the environment.
- Waste disposal of the polymers [7].
- Uses of polymers in medical application and transportation such as automobiles, aviation industry, etc.

8.2 Synthesis of Polyethylene

Synthesis of polyethylene assemblage from ethylene mostly involves step-step protocol under various conditions of temperature and pressure in a reactor, a wide range of polymers may be made both softer and harder material [7]. Polyethylene comes in many forms such as high density, low density, linear, hyper-branched. The polymerization of ethylene release by radical initiators. Radical initiators give more or less branched polymer chains.

$$nCH_2{=}CH_2 \longrightarrow \left[-CH_2-CH_2-\right]_n$$

Ethylene Polyethylene

Both low- and high-molecular-weight polymers synthesized by either organometallic coordination or high-pressure radical polymerization. The structure of the polyethylene differs with the two methods. Radical initiators can carry out the ethylene polymerization into polyethylene. Polyethylene is a simple chemical reaction of chain growth polymerization. Oxidative deterioration is center of sensitivity of chemical attack.

8.3 Polypropylene

Atactic polypropylene can be isolated without the use of Ziegler – Natta catalysts or homogeneous catalysts. The polymer can be isolated through extraction techniques with n – hexane or n – heptane [8].

$$nCH_2{=}CH(CH_3) \longrightarrow \left[-CH_2-CH(CH_3)-\right]_n$$

Propylene Polypropylene

8.3.1 Synthesis of Polystyrene

Synthesis of polystyrene from styrene monomers is capable of undergoing carbocationic polymerization. The double bond of styrene acts either as electron donating or as electron withdrawing center [9–11]. They are able to form stable carbocation due to the resonance stabilization of cation over aromatic ring.

$$\mathrm{HC{=}CH_2}\text{-}C_6H_5 \xrightarrow{H^+} \overset{\oplus}{\mathrm{HC}}\mathrm{-CH_3} \longleftrightarrow \mathrm{HC-CH_3} \longleftrightarrow \mathrm{HC-CH_3} \longleftrightarrow \mathrm{HC-CH_3}$$

$$n\ \mathrm{CH_2{=}CH(C_6H_5)} \longrightarrow \mathrm{-[CH_2-CH(C_6H_5)]_n-}$$

Styrene Polystyrene

Synthesis of polystyrene can also be synthesized using different initiators in combination with various Lewis acids such as BCl_3, $SnCl_4$, $TiCl_4$, $AlCl_3$, etc., in solvents [12–16].

8.4 Synthesis of Polyamide

8.4.1 Synthesis of Nylon 6

Nylon 6 has many applications such as textile fiber, which it finds utilization in textile carpeting materials. Simple synthesis starts from ammonia and acetic acid with the formation of acetamide as well as ethylamine and acetic acid with formation of *N*- ethylacetamide.

$$\underset{\text{Ammonia}}{NH_3} + \underset{\text{Acetic acid}}{CH_3COOH} \longrightarrow \underset{\text{Acetamide}}{NH_2COCH_3} + H_2O$$

$$\underset{\text{Ethylamine}}{C_2H_5NH_2} + \underset{\text{acetic acid}}{CH_3COOH} \longrightarrow \underset{\text{N - ethylacetamide}}{C_2H_5NHCOCH_3} + \underset{\text{water}}{H_2O}$$

However, amino groups combine with carboxyl groups on the same molecule form linear polymer. The carboxyl group will go on one molecule further react with amino group on another. The process goes on, the molecules condensing together to give longer and longer molecules. These stepwise syntheses are time and cost consuming. It has been made to obtain materials with compatible properties through shorter synthesis routes.

$$2\,NH_2(CH_2)_5COOH \longrightarrow NH_2(CH_2)_5COO\,NH(CH_2)_5COOH$$

ε - aminocaproic acid — dimer

$$n\,NH_2(CH_2)_5COOH \longrightarrow H{-}\!\left[NH(CH_2)_5CO\right]_n\!{-}OH$$

ε - aminocaproic acid — Nylon 6

Synthesis of nylon by ring opening polymerization of ε-caprolactam is one of the important polymerization. Synthesis and technical production of polyamide-6 is the formation of the polyamide chains is accompanied by the formation of cyclic oligoamides Polymerization of ε-caprolactam with formation of 6 – aminohexanoic acid (from hydrolysis of ε-caprolactum) that on heating to 260°C [17–19].

n ε-caprolactam + H_2O ⟶ n 6 - Aminohexonoic acid

ε-caprolactam

6 - Aminohexonoic acid

n 6 - Aminohexonoic acid $\xrightarrow[-nH_2O]{260^oC}$ Nylon 6

6 - Aminohexonoic acid

Nylon 6

8.4.2 Synthesis of Nylon 6,6

Nylon 6,6 is an aliphatic polyamide. Nylon 6,6 also called as poly(hexamethyleneadipamide). It is synthesized using step growth polymerization. The monomers involved are hexamethylenediamine and adipic acid which is bifunctional with elimination of water. While synthesis going on the reaction mixture contains wide distribution of slowly growing oligomers forming at one end with hydroxyl and other end with amino groups. Nylon 6,6. Heating of the salt of adipic acid and hexamethylenediamine to 280°C (high purity to obtain long chains).

$$n\,HO-\overset{\overset{O}{\|}}{C}-(CH_2)_4-\overset{\overset{O}{\|}}{C}-OH + n\,H-NH(CH_2)_6NH-H$$

Adipic acid — Hexamethylenediamine

$$\longrightarrow \left[\overset{\overset{O}{\|}}{C}-(CH_2)_4-\overset{\overset{O}{\|}}{C}-NH(CH_2)_6NH\right]_n$$

Nylon 6,6

In other synthesis by using adipoyl chloride, the reaction starts with molecule of hexamethylenediamine. The hydrogen atom of the amine group from hexamethylenediamine forms hydrochloric acid molecule with chloride from the acid chloride functional group. The remaining molecules of both monomers will join to from polymer. The nylon 6,6 molecule formed has an acid chloride group at one end and an amine group at the other end.

$$n\,H_2N-(CH_2)_6-NH_2 + n\,HOOC-(CH_2)_4-COOH$$

Hexamethylene diamine — Adipic acid

$$\xrightarrow{-H_2O} H\left[HN-(CH_2)_6-NH-OC-(CH_2)_4-CO\right]_nOH$$

Nylon 6,6

$$n\,H_2N-(CH_2)_6-NH_2 + n\,ClOC-(CH_2)_4-COCl$$

Hexamethylene diamine — Adipoyl chloride

$$\xrightarrow{-HCl} H\left[HN-(CH_2)_6-NH-OC-(CH_2)_4-CO\right]_nCl$$

Nylon 6,6

8.4.3 Synthesis of Polyethylene Terephthalate

Polyethylene terephthalate synthesized from diacid, diol and also low molecular weight polymers to form cross-linked polymers. Synthesis of polyethylene terephthalate involves the reaction between terephthalic acid and ethane

1,2 diol with the elimination of water molecules by esterification reaction. It has mechanical properties due to its aromatic group provides rigidity. It has toughness due to its flexible backbone and its partial crystallinity. The solubility of terephthalic acid in ethylene glycol to form polyethylene terephthalate is very low [20, 21]. The solubility of one of the monomers in step growth polymerization is limited. Poly(ethylene terephthalate) (PET) reactions occur between ethylene glycol and dimethyl terephthalate.

$$nHOOC-C_6H_4-COOH + nHOCH_2-CH_2OH$$

Terephthalic acid ethane 1,2 diol

$$\xrightarrow{-2n\,H_2O} \left[-OCH_2-CH_2-OOC-C_6H_4-CO- \right]_n$$

Polyethylene terephthalate

A polyethylene terephthalate proceeds by ester interchange between the hydroxyl ends group and the ester linkage. Reaction proceeds by ester-ester linkage between two ethylene diester groups situated along polymer chains. Linking of the ends of two chain molecule is carried out by the elimination of ethylene glycol [22]. Synthesis of polyethylene terephthalate undergoes with functional groups such as carboxylic acid, ester or acid chloride, and hydroxyl such as 1,2 ethandiol (glycol), 1,2 propandiol, 1,2,3 propantriol (glycerol).

$$n\,CH_3OOC-C_6H_4-COOCH_3 + nHOCH_2-CH_2OH$$

Dimethyl terephthalate ethane 1,2 diol

$$\xrightarrow{-2nCH_3OH} \left[-OCH_2-CH_2-OOC-C_6H_4-CO- \right]_n$$

Polyethylene terephthalate

Polyethylene terephthalate synthesis also involves reaction between dimethyl terephthalate and ethane 1,2 diol (ethylene glycol) with reaction mixture of slowly growing oligomers. Synthesis of polyester with diacid needs high temperature whereas with diester, the polymerization requires lower temperature [20, 21].

8.5 Polystyrene

The first step of the polystyrene synthesis is a Diels – Alder reaction with formation of adduct of monomers.

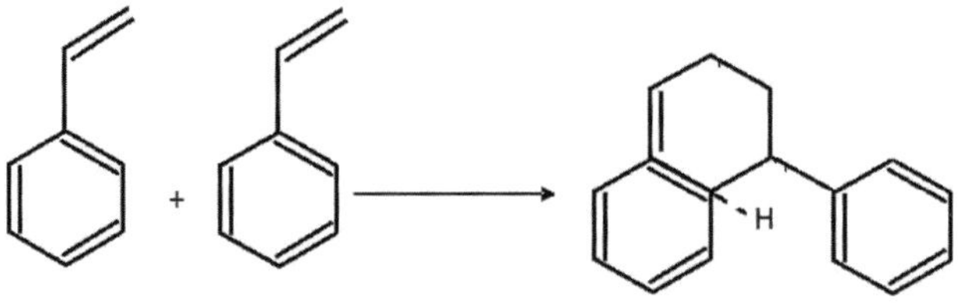

+ H

Styrene polymerizes without a chemical initiator by heating. The reaction is spontaneous polymerization. Styrene further react with adduct which yields two radicals to undergo polymerization. Some of the dimers and trimer present,

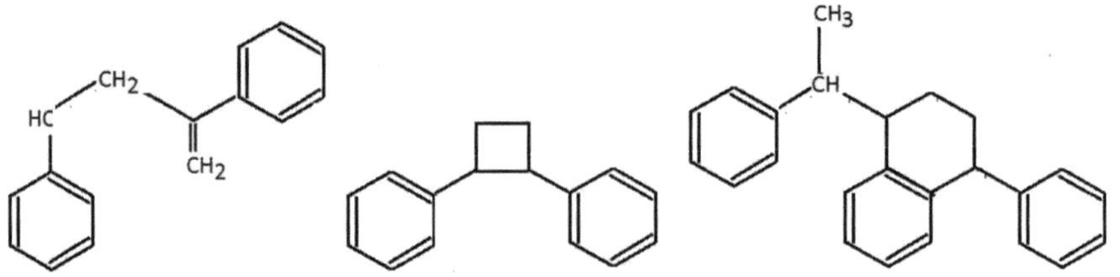

CH_3 CH_2 HC CH_2 CH

also have been identified [23, 24]. This will be an impurity in the final product.

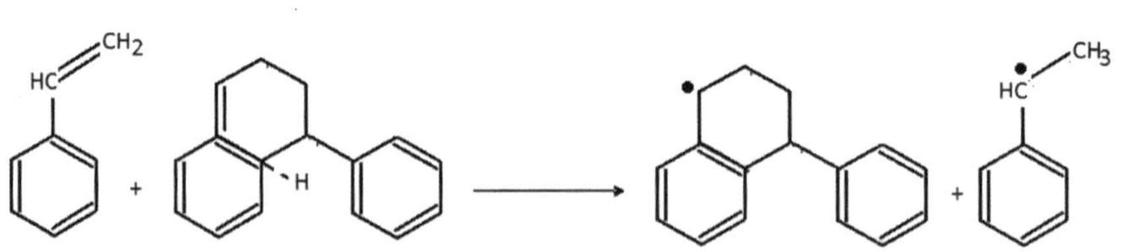

CH_2 HC + H CH_3 HC +

In polystyrene, the styrene terminates by the condensation of two growing chains in free radical polymerization [25, 26]. Termination is diffusion controlled at all temperatures below 150°C [27, 28]. Increasing viscosity leads to a reduction in the termination rate [29]. However, the resulting Trommsdorff effect is comparably small for polystyrene [30].

8.6 Polyvinylacetate

Vinyl acetate manufacture is by the reaction of ethylene with acetic acid in present of palladium catalyst is the dominant method of commercial production. Synthesis by free radical polymerization, vinyl acetate forms both head-to-head and head-to-tail polyvinylacetate addition can take place [31]. The reaction results in the incorporation of the two types of repeating units in the backbone of the polymer [32]. The proportion of these two types is dependent on the temperature at which the polymerization carried out. With an increase in polymerization temperature, preferably higher content of head-to-head obtains.

$$H_3C-COO-CH=CH_2 \longrightarrow \sim CH_2-CH(OAc)-CH_2-CH(OAc)\sim + \sim CH_2-CH(OAc)-CH(OAc)-CH_2\sim$$

8.7 Polybutadiene

Synthesis of syndiotactic polybutadiene is a crystalline polymer from the polymerization of butadiene with compounds of titanium, cobalt, vanadium and chromium [33, 38], alcoholates such as cobalt (II) 2-ethylhexanoate, titanium (III) butanolate with triethylamine as cocatalyst. Amorphous polybutadiene produced with molybdenum (V) chloride and diethylmethoxyalumium. Addition of esters of carboxylic acids raises the vinyl content of the product. Trisallylchromium polymerizes 1,3-butadiene to 1,2-polybutadiene, while bisallylchromchloride gives 1,4-polybutadiene [39].

n $CH_2{=}CH{-}CH{=}CH_2$ $\xrightarrow{(\text{allyl})_3\text{Cr}}$ $-[CH_2-CH(CH{=}CH_2)]_n-$

n $CH_2{=}CH{-}CH{=}CH_2$ $\xrightarrow{(\text{allyl})_2\text{Cr–Cl}}$ $-[CH_2-CH{=}CH-CH_2]_n-$

8.8 Polyisobutylene

Isobutylene undergo polymerization by cationic polymerization in presence of $AlCl_3 - CH_3Cl$ catalyst. In the cationic polymerization of isobutylene, one of the rare cases of an intermolecular carbenium ion addition to an olefin without polymer formation occurs in the industrial synthesis of isooctane.

$(H_3C)_2C{=}CH_2 + H^+ \longrightarrow (H_3C)_2C^{\oplus}{-}CH_3$

$(H_3C)_2C{=}CH_2 + (H_3C)_2C^{\oplus}{-}CH_3 \longrightarrow (H_3C)_3C{-}C^{\oplus}(CH_3)_2$ (CH_2)

$(H_3C)_3C{-}C(CH_3)_2{-}CH_2^{\oplus} + (H_3C)_2C^{\oplus}{-}CH_3 \longrightarrow H_3C{-}C(CH_3)_2{-}CH_2{-}C(CH_3)_2{-}CH_2{-}C(CH_3)_2{-}CH_2^{\oplus}$

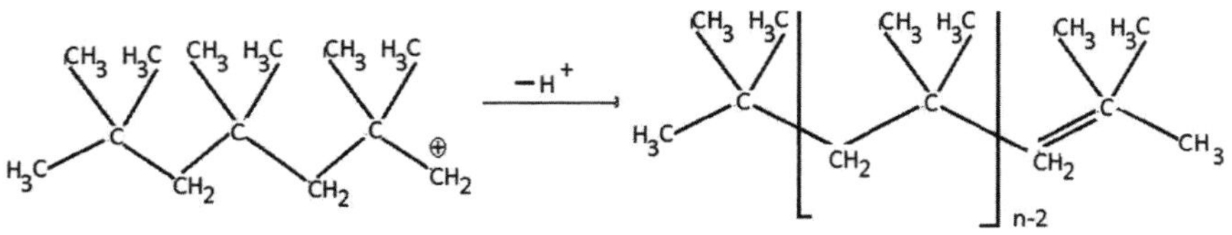

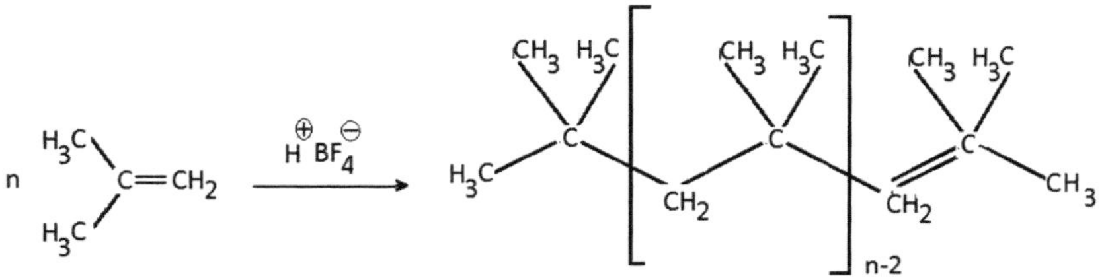

The order on the yield of polymer from the polymerization of isobutylene by Friedel – Crafts Catalysts is

$$BF_3 > AlBr_3 > TiCl_4 \geq TiBr_4 > SnCl_4 > BCl_3 \text{ or } BBr_3$$

Intermolecular additions of carbenium ions to olefins give polymers. Such a reaction used in industry. Carbenium ion additions is to isobutylene as key steps in the cationic polymerization of isobutylene and the mechanism with additions that take place or can take place without stereocontrol [40, 41].

8.9 Polymethylmethacrylate

At low temperature, combination and disproportionation are important in methyl methacrylate polymerization. It has dominant role in the polymerization [42–44]. Termination reactions are fast and difficult to control due to bimolecular termination involvement with radical centers on polymeric reactants. Auto-acceleration of polymerization is important in terms of diffusion-controlled termination. The large increase in viscosity of the medium, gives an increase in radical concentration. This effect of the auto-acceleration is called gel effect or Trommsdorff-Norrish effect [45].

Polymerization of acrylic esters, i.e. methyl acrylate, ethyl acrylate, butyl acrylate, and tert-butyl acrylate, initiated by rare earth metal complexes were non-stereospecific [46].

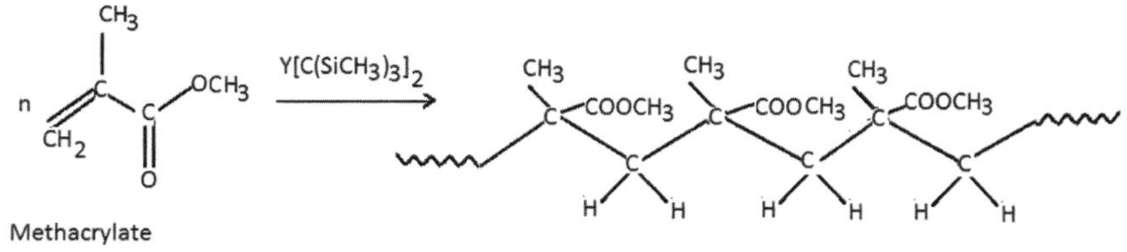

Methacrylate

Polymethylmethacrylate

Methylmethacrylate is polymerized into polymethylmethacrylate using iodo-malonates in combine with $(nBu)_2N^+I^-$ as inhibitors which is new initiator system and specific for methacrylate and not polymerize acrylates [47].

$(nBu)_4\overset{\oplus}{N}\overset{\ominus}{I}$, Warm

8.9.1 Synthesis of Polyurethane

Synthesis of polyurethane starts with the reaction of a compound contains at least two isocyanates with other polyfunctional reagents containing active hydrogen. Addition of isocyanate with diol is a spontaneous reaction with liberation of heat. Polyurethane synthesis between diisocyanate and diol leads to the formation urethane. Linear polyurethane synthesized from 1,4-butanediol and hexamethylene diisocyanate.

$$O{=}C{=}N{-}(CH_2)_6{-}N{=}C{=}O + HO{-}(CH_2)_4{-}OH \longrightarrow \left[-O{-}(CH_2)_4{-}OCNH{-}(CH_2)_6{-}NHCO- \right]_n$$

Hexamethylene diisocyanate — Butane diol — Polyurethane

The reaction of a compound contains at least two isocyanates with other polyfunctional reagents containing active hydrogen. Linear polyurethane synthesized from diols such as 1,4 – butadiol and hexamethylene diisocyanate are usually made in aromatic chlorinated solvents at reflux. Polyurethane synthesized from hexamethylene diisocyanate and cis-and trans-2 butene 1,4, diol show small differences in properties, not attributable to cis-trans isomerism [48, 49]. Polymers synthesized from 1,3 propane diol and 2,2-dimethyl-1,3-propanediol by reaction with hexamethylene diisocyanates are more crystalline than is that from 2-methyl-1,3-propanediol [50].

8.9.2 Electro-Polymerization of Polypyrrole

Electrochemical oxidation of pyrrole forms a film of conducting polymer at the electrode surface. Initiation and formation of monomer radical cation is done by electrochemical oxidation. Combination of two radical cation monomers (or oligomers) followed by loss of two hydrogen ions. The linkage formed is at the 2 position of the pyrrole ring, forming 2,2'-bipyrrole. 2,5-disubstituted pyrroles do not polymerize and 2-monosubstituted pyrroles only form dimmers and further by re-oxidation of the bipyrrole and further combination of radicals.

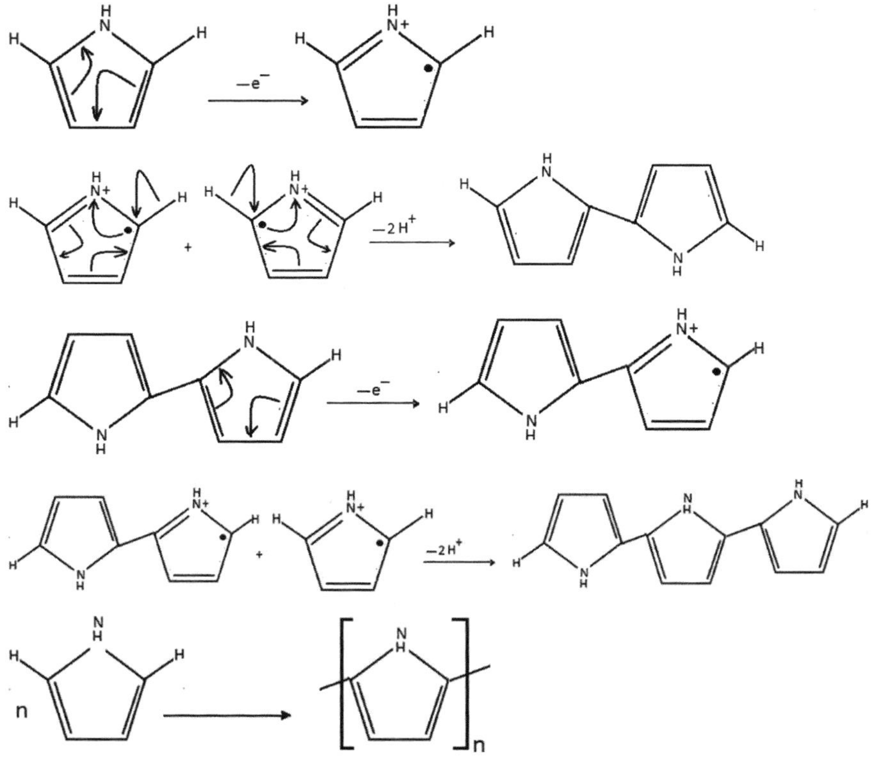

The electropolymerization stops no further monomer is present for oxidative polymerization or side reactions terminate the polypyrrole chain. The success of electropolymerization of pyrrole is due to the stability of the radical through charge delocalization, and the ease of electro-oxidation. The loss of the hydrogen ions makes the dimer (oligomer) formation irreversible so proton acceptors, such as water, pyridine and bases, enhance electropolymerization. Good solvents for electropolymerization include water, acetonitrile, butanone, propylene carbonate, dimethylformamide (DMF) and ethanol though the presence of a bit of water enhances the polymer formation. Water can also result in chain termination. Pyrrole is to polymerize to give conducting powder in black color. In doped form, it has better chemical and thermal stability. In neutral form, polypyrrole films are yellow/green and are sensitive to air and oxygen [51, 52].

8.9.3 Synthesis of Polylactide (PLA)

Synthesis of polylactide from polycondensation of lactic acid is an economic process due to its polymerization in liquid phase. The process requires temperature increase to remove the moisture. Therefore, it is impossible to synthesise polylactide at high temperature due to its decomposition. A high molecular weight polylactide (PLA) is synthesized by ring opening polymerization of cyclic ester of lactic acid.

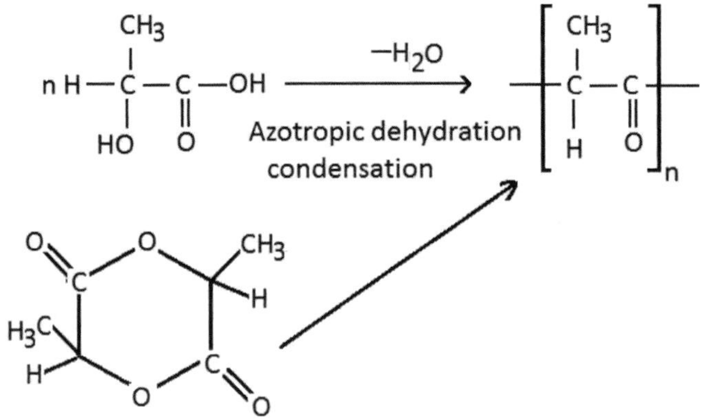

$-H_2O$

Azotropic dehydration condensation

The process is very complex and the economy of the process is high. It is particularly green polymer in terms of sustainability and degradation in the life cycle analysis. The polymer is formed either by direct condensation of lactic acid or cyclic intermediate dimer namely lactide through a catalysed ring-opening polymerization process [53–55]. Poly (D,L – lactide) is amorphous with poor drug permeability but much higher degradation rate.

8.9.4 Name Reactions and Polymer Synthesis

Polymer chemistry is a subject of its impressive way to construct the macromolecular structures. Therefore, with rapid development, the name reaction has transferred from small molecules to polymers. Name reactions have played an enormously significant role in shaping chemical synthesis. Even synthetic methods developed for polymers too. Many of the reactions are constantly and repeatedly under examination until to attain new applications.

8.10 Reppe Chemistry

8.10.1 Poly(*N* – vinyl carbazole)

Reppe chemistry is particularly important to the manufacture of many high polymers and other synthetic products [56]. Acetylene chemistry is involving the use of acetylene at high pressures in the presence of suitable catalysts to carry out the fundamental reactions of vinylation, ethynylation, cyclopolymerization and carbonylation [57–62]. Poly (N – vinyl carbazole) is a vinyl aromatic polymer. Synthesis of poly (N – vinyl carbazole) is from N – vinyl carbazole which is controlled by the electronic and steric hindrance of the carbazole group [63]. The monomer and polymer reaction and the structure and properties controlled by the electronic and steric hindrance of the carbazole group.

$CH_2{=}CH$ N-vinyl-carbazole → $-[CH_2-CH]-$ Poly(N-vinylcarbazole)

8.11 Michael Adducts

8.11.1 Derivative of *N*-Vinylformamide and Polymer Synthesis

Michael adducts reaction shows chemistry for synthesis of new derivatives by reaction. The derivatives exhibit facile free radical reactivity in polymerization. N-Vinylformamide, is utilized as a reactive molecule for adduct formation with ester. It is water soluble to cationic as precursors and reactive amine functional polymers. The polymers show good thermal, chemical stability, and very low viscosity [64–66].

Base cat.

N - Vinylformamide

Michael adduct

N - Vinylformamide

NaH

Polymer derivative

N - Vinylformamide

Michael adduct

Tin Catalyst

Michael adduct

Diisocyanate

Polymer

8.12 Polymer Synthesis – Future

The polymerization and chemistry elaborate to provide unique method synthesis. They are simple, applicable and suitable for specialty polymers. The scientific curiosity is from the practical and profit of synthesizing of polymers. Polymer synthesized with required properties with definite application accompanies with the development of new analytical techniques with capabilities.

Explosive growth of research into the polymer synthesis is the primary reason for the great interest in polymeric materials lies in the major changes that can achieved in the properties of polymers. Still some of the areas in polymer syntheses are in their infancy. Emerging technologies in chemistry, engineering, medicine and biotechnology, the polymers are for potential applications and versatile as new dimension in polymer chemistry.

In future, polymer synthesis is greatly facilitated by radicals, catalysts, etc., further on by catalyst, solvent recovery, and recycle. Catalyst and solvent reuse increases the overall productivity and cost effectiveness of polymerization reactions while minimizing their environmental impact. Ultimately it is contributing considerably to the sustainability of synthetic polymer processes. In the case of non-biodegradable polymers, the only criterion for the quality of a catalyst or initiator is its performance. However, the shortcomings of many synthetic polymers lie in sequences, often leads to structure augmentation.

Therefore, a high performance polymer synthesis defined by the following points:

(A) High productivity, which means rapid polymerizations and large quantities of polymers per molecule of catalyst;
(B) High molar masses should be accessible;
(C) Initiator or catalyst allows for tailor making the molar masses;
(D) Narrow molar mass distributions are desirable for syntheses of polymers.
(E) Various architectures should be accessible (e.g., random copolymers, block-copolymers, star-shaped polymers, etc.).
(F) Suppression of Side reactions.

Innovation of the synthesis of polymeric materials with required properties is the curiosity of scientific, practical and profit. Innovation in the synthesis of polymeric materials should accompany with development of new analytical techniques with capabilities.

8.13 Summary

- Polymer synthesis represents the mechanics of chemical connections that form giant molecules.
- Chemistry is to obtain a molecular level and coherent understand about the polymer synthesis.
- The laboratory synthesis as a test-bed for assessing the utility and importance of polymer synthesis.
- Polymer synthesis must be concerned not only with the cost-effective production but also with the environmental impact of the reactions.

- Polymer synthesis is in their early stages of development, traditional, but methods that are more efficient to elaborate and continue. Elaboration of the synthesis has to be addressed a more general and often rather troublesome strategic task.
- The growing polymer synthesis in modern multidiscipline research ranging from commercial to industrial.
- The pathway of polymer synthesis includes realization of each step may represent independent chemical problem.
- New synthetic methods have emerged from the deliberate attempt to perfect a known reaction or invert a new one in order to permit the preparation of a specific target polymer.

References

[1] E. J. Corey, and X. Cheng. *The Logic of Chemical Synthesis*. New York, NY: Wiley-Interscience. (1989).
[2] I. Fleming. *Selected Organic Syntheses*. Wiley: London, (1973).
[3] K. C. Nicolaou, and E. J. Sorensen, and N. Winssinger. *J. Chem. Educ.*, **75**, 1225–1258 (1998).
[4] E. J. Corey. *Pure Appl. Chem.* **14**, 19–37 (1967).
[5] A. van der Horst, and P. J. Schoenmakers. *J. Chromatogr. A*, **1000**, 693–709 (2003).
[6] N.G. Kumar. *Macromol. Rev.*, **15**, 225 (1980).
[7] X. Zhang, M. F. A. Goosen, U. P. Wyss, D. and Pichora. *J. Macromol. Sci. Rev. Macromol. Chem. Phys.*, C33 (1993).
[8] W. Marconi, S. Cesca, and G. Della Fortuna. *Chim. Ind.*, **46**, 1131 (1964).
[9] J. P. Kennedy, and E. Marechal. *Carbocationic Polymerization*. Wiley-Interscience: New York (1982).
[10] Y. Ishihama, M. Sawamoto, and T. Higashimura. *Polym. Bull.*, **23**, 361–366 (1990).
[11] Z. Fodor, M. Gyor, H. C. Wang, and R. J. Faust. *Macromol. Sci. Pure Appl. Chem.*, **30**, 349 (1993).
[12] M. R. Ribeiro, M. F. Portela, A. Deffieux, H. Cramail, and M. Rocha. *Macromol. Rapid Commun.*, **17**, 461–469 (1996).
[13] T. Hasebe, M. Kamigaito, and M. Sawamoto. *Macromolecules* **29**, 6100–6103 (1996).
[14] K. Satoh, J. Nakashima, and M. Kamigaito. *Macromolecules*, **34**, 396–401 (2001).

[15] J. M. Oh, S. J. Kang, O. S. Kwon, and S. K. Choi. *Macromolecules*, **28**, 3015–3021 (1995).
[16] Y. Deng, C. Peng, P. Liu, J. Lu, and L. J. Zeng. *Macromol. Sci. Pure Appl. Chem.*, **33**, 995 (1996).
[17] S. K. Gupta, and A. Kumar. *Polymer*, **22**, 481 (1981).
[18] V. S. Kumar, and S. K. Gupta. *Ind. Eng. Chem. Res.*, **36**, 1202 (1977).
[19] D. Heikens. *Polymer*, **22**, 1758 (1981).
[20] I. Goodman. "Direct esterification process," in *Encyclopedia of Polymer Science and Technology*, 2nd Edn. eds Mark, H. F., and Kroschmitz, J. I. (Wiley-Interscience: New York, NY), **12**, 45–46 (1988).
[21] Poly (1980). Ethylene terephthalate. Japanese Patent 55,112,232.
[22] G. Challa. *Makromolekulare Chem.*, **38**, 105 (1960).
[23] O. F. Olaj, H. F. Kauffman, and J. W. Breitenbach. *Makromol Chem.*, **178**, 2707 (1977).
[24] D. F. Stein, and H. Mosthaf. *Angew. Makromol. Chem.*, **2**, 39 (1968).
[25] J. C. Bevington, H. W. Melville, and R. P. Taylor. *J. Polym. Sci.*, **14**, 463 (1954).
[26] G. Ayrey, F. G. Levitt, and R. J. Mazza. *Polymer*, **6,** 157 (1965).
[27] G. Weickert, and R. Thiele. *Plaste Kautschuk*, **8**, 432 (1983).
[28] I. Mita, and K. Horie. *Rev. Macromol. Chem. Phys. C*, **27**, 91 (1987).
[29] T. J. Tulig, and M. Tirrell. *Macromolecules*, **14**, 1501 (1981).
[30] Dow Family of Styrenic Monomers *Technical Bulletin 11 5-608-85*. Midland, MI: The Dow Chemical Company (1985).
[31] W. A. Mowers, and J. V. Crivello. *Polym. Mater. Sci. Eng.*, **81**, 479 (1999).
[32] Flory, P. J., and Leutner, F. S. (1948). J. Polym. Sci., 3: 880; (1950). 5: 267.
[33] G. Natta, L. Porri, and G. Mazanti. (1955). DOS 1420553 to Montecatini, C. A. (1959) **53**: 3756.
[34] G. Natta, L. Porri, and A. Carbonaro. *Makromol. Chem.*, **77**, 207 (1964).
[35] M. Ichikawa, H. Kurita, and A. Kogure. (1966). US.P. 3498963 to Japan Synthetic Rubber, C.A. (1969) **71**, 82394.
[36] W. Kampf, and K. Nordsiek. DOS 2447203 to Chemische Werke Huels, C.A. **85**, 47938 (1976).
[37] S. Sugiura, H. Ueno, and H. Hamada. US.P. 3778424 to Ube Ind., C.A. (1971) **75**, 65081 (1970).
[38] H. Ashitaka, H. Ishikawa, H. Ueno, and A. Nagasaka. *J. Polym. Sci., Polym. Chem Ed.*, **21**, 1853 (1983).

[39] F. Dawans, and P. Teyssie. US.P. 3451987 to Institut Francais du Petrole, C.A. (1968) **68**, 3663 (1965).
[40] J. P. Kennedy and R. M. Thomas. *Polym. Sci.*, **46**, 481 (1960).
[41] J. P. Kennedy and R. M. Thomas. *Polym. Sci.*, **46**, 233 (1960).
[42] G. V. Schulz, G. Henrici-Olive, S. Olive. *Makromol. Chem.* **31**, 88 (1959).
[43] C. H. Bamford, G. C. Eastmond, D. Whittle. *Polymer* **10**, 771 (1969).
[44] M. Stickler, D. Panke, and A. E. Hamielec. *J. Polym. Sci., Polym. Chem. Ed.*, **22,** 2243 (1984).
[45] G.V. Schulz, G. Haborth. *Makromol. Chem.*, **1**, 106 (1948).
[46] K. Soga, H. Deng, and T. Shiono. *Macromolecules*, **28**, 3067 (1995).
[47] O. W. Webster, W. R. Hertler, D. Y. Sogah, W. B. Farnham, and T. V. Rajan Babu. *J. Am. Chem. Soc.*, **105**, 5706 (1983).
[48] O. Bayer. *Modern Plast.*, **24**, 250 (1947).
[49] C. S. Marvel, and C. H. Young. *J Am Chem Soc.*, **73**, 1066 (1951).
[50] W. Brenschede. German Patent **766**:888 (1953).
[51] G. P. Gardini. *Adu. Heterocycl. Chem.*, **15**, 67 (1973).
[52] J. A. Walker, L. F. Warren, and E. F. Witucki. *J. Polym. Sci.*, **26**, 1285 (1988).
[53] R. E. Drumright, P. R. Gruber, D. E. Henton. *Adv. Mater.*, **12**, 1841 (2000).
[54] R. A. Gross, and B. Kalra. *Science*, **297**, 803 (2002).
[55] H. Tsuji, and Y. Ikada. *J. Polym. Sci. A*, **36**, 59 (1998).
[56] P. J. Stang, and Diederich F. *Modern Acetylene Chemistry*, (John Wiley: New York, NY) 1995.
[57] J. W. Copenhaver, and M. H. Bigelow. *Acetylene and Carbon Monoxide Chemistry,* New York, NY, 246 (1949).
[58] D. W. F. Hardie. *Acetylene, Manufacture and Uses* (New York, NY) 67 (1965).
[59] L. F. Fieser, M. Fieser. *Reagents for Organic Synthesis* (New York, NY) 61 (1967).
[60] A. Mullen. "Carbonylations catalyzed by metal carbonyls-reppe reactions," in *New Syntheses with Carbon Monoxide*, Ed. Falbe, J. (Springer-Verlag, Berlin) 243–308, (1980).
[61] R. E. Colborn, and K. P. C. Vollhardt. Mechanistic study of cyclooctatetraene synthesis. *J. Am. Chem. Soc.*, **108**, 5470 (1986).
[62] C. J. Lawrie et al., *Organometallics* **8**, 2274 (1989).
[63] W. Reppe, and E. Keyssner. Ger. 618, 120 to IG Farbenindustrie; C.A. (1936), **30**, 110 (1935).

[64] R. K. Pinschmidt, Jr. W. L. Renz, W. E. Carroll, K. Yacoub, J. N. Drescher, A. F. Nordquist, and N. Chen. *J. Macromol. Sci. Pure Appl. Chem.*, **A34**, 1885–1905 (1997).
[65] R. K. Pinschmidt, Jr. and N. Chen *Polym. Preprints*, **39**, 639–640 (1998).
[66] K. Pinschmidt, Jr. and D. J. Sagl in Polymeric Materials Encyclopedia, J. C. Salamone, ed.-in-chief, CRC Press: New York, NY, 7095–7103 (1996).

9

Polymerization Processes

In modern organic polymer chemistry, step- and chain-growth polymerization methods are critically important for the fabrication of high-volume and specialty plastics, foams, gels, and rubbery materials. Such materials have enabled new applications that have transformed the society [1, 2]. Polymerization is concerned with reactions of monomers. The high molecular weight polymers are products especially with the nature of polymerization reactions.

The monomer combines with the operation of primary valence forces. Many compounds are not unsaturated and are capable of reacting themselves to form polymers. However, the polymerization process is peculiar to unsaturated compounds with polycondensation or polyaddition process. Therefore, polymerization processes must be enabled to have the required product characteristics with respect to the demand of the market at economical costs. The continuous polymerization processes become more and more important for production of the consumer products.

9.1 Essential Ingredients in Polymerization Process

Apart from the monomer, different solvents are used which may be free from minor residues or purification if needed. The impurities present may influence the extent type of side reactions. Therefore, the monomer and solvents which are used in polymerization include a source, specific identification as safety and pure.

The effects of initiators, solvents, chain transfer agents, and inhibitors on the mechanisms of polymerization is an important phenomena due to the growing polymer which may change dramatically and the rate coefficient may vary significantly upon the addition of a single molecule of the monomer. These changes would reflect in the polymer properties. The other ingredients which are essential in polymerization process as follows:

9.1.1 Initiators

An initiator is an important ingredient to start many of the polymerizations particularly radical polymerization. These initiators decompose and undergo initiation of process. Many initiators provide opportunity for changing the molecular weight and distribution [3–6]. Initiators undergo decomposition and polymerization reactions according to their half-life times [7–10]. The molecular weight modification is by the use of chain transfer agents such as hydrocarbons, alcohols, aldehydes, ketones, and esters [11, 12].

Some of the polymerization initiators are:

- With thermal decomposition organic compounds such as 2,2-azobisisobutyronitrile(AIBN), dibenzoyl peroxide
- Bimolecular initiating system such as redox initiation of Fe^{3+} and Ce^{4+}
- Boron alkyls and metal alkyl initiators such as dialkylzinc, dialkylcadmium, triethylaluminum, etc.
- Ultra violet as photochemical initiator
- Radioactive sources and electron beams such as gamma rays, beta rays, alpha particles.
- Ultrasonic radiation.

9.1.2 Surfactants

Surfactants are important for latex particles nucleation, monomer droplets emulsification, and/or preformed polymer, and stabilization of polymer particles during polymerization or shelf life of the products. These can have adverse effects such as foam formation [13, 14]. During mixing with other products in formulation, surfactant can migrate. Therefore, destabilization of the latex particles occur [14].

9.1.3 Catalysts

A catalyst used in the polymerization processes should have the following main characteristics:

- Activity to reach high conversion of the polymer
- Removal of catalyst residues from the polymer without any trouble.
- Provide resulting polymer with suitable molecular weight distribution
- Provide high level of molecular weight with the maximum stereo regularity degree of polymerization phase
- Without any need to remove the fractions of irregular polymer

- Capable to obtain a granular polymer with proper morphology, particle size distribution, and compactness both in manufacturing and product utilization.

The developments in both Ziegler–Natta (ZN) and metallocene (Me) type catalysts has allowed increasing control on the type, amount and distribution of the α-olefin comonomer within the chains. The knowledge of the distribution of the α-olefin along the chain is particularly important, since many of these copolymers can exhibit a highly heterogeneous intra- and/or inter-comonomer distribution.

9.1.3.1 Ziegler–Natta catalyst

Ziegler–Natta catalysts are highly sensitive to oxygen, moisture, and a large number of chemical compounds. Reagent purity and catalysts manipulations are important for achieving reproducibility and reliability in polymerization. Requirement of pure solvents and monomers substantially improves polymerization. Alkanes and aromatic compounds have no substantial effect and are used as solvents which cannot affect the polymerization. The mechanism of this catalyst is not well-known. A monomer is absorbed at the transition metal in an active complex. Activated monomer inserts into the metal-carbon bond. The metal-organic polymerization resembles similar to nature accomplished with enzymes. The Ziegler catalyst (triethyl aluminum-titanium tetrachloride) for the polymerization of ethylene at low pressures to essentially linear polymers.

9.1.3.2 Phillips catalyst

Philips catalyst is alkyl free. It can be prepared by impregnating chemical compounds with high surface supports are used. It contains chromium oxide along with the impregnated compounds. The powder catalyst fluidizes and activates by a stream of dry air to remove bound water. Philips catalyst catalyzes the polymerization produces polymer with lower crystallinity. The chromium oxide is inactive at lower temperatures and becomes active as the temperature increases. Phillips catalyst has using hexavalent chromium oxide supported on silica-alumina claimed to be similar to the Ziegler material but not identical [15–18].

9.1.3.3 Metallocene catalyst

Metallocene catalyst is used not only for new polyethylenes and polypropylenes, but also for more polyolefins that are complex. Metallocene catalysis is the molecular mechanism and the origin of stereoselectivity in R – olefin polymerization catalysis [19, 20]. Stereo control during propylene polymerization has become possible through the rational design of the metallocene structure [21, 22]. However, there is no general method to influence the stereoselectivity during the polymerization of the vinyl aromatic commodity monomer styrene [23–25]. They are suitable for synthesizing cyclic olefin copolymers (COCs) from ethylene and norbornene.

9.1.4 Chain Transfer Agents

Chain transfer agents can be solvent, monomer, initiator, polymer, or an added chemical agent. It effects in molecular weight and branched structures formation in the polymer. The most important aspect of chain transfer is the control of molecular weight by the adequate use of added transfer agent. Mercaptans are the most widely used chemicals as inhibitor [26, 27].

Possible effects of chain transfer agent (CTA) are [28]:

- Reduces molecular weight, bulk viscosity, and enhances monomer diffusivity.
- Reduces gel formation by limiting successive chain transfer to polymer reactions.
- Growing polymer branch is much more likely to terminate before it can form a network.

9.1.5 Solvents

The solvents are important in every polymerization reaction. They can influence the stereo-regularity of the product, particularly in anionic polymerizations. Solvent boiling point should correspond to that of the monomers and to the decomposition temperature of the initiators [29].

9.1.6 Suspending Agents

Phase ratio, temperature, monomer addition, and other additives have effects on suspension stability. Suspending agents are important and critical art. Two classes of suspension stabilizers are [30, 31]:

9.1.6.1 Water-soluble polymeric compounds

Water-soluble polymeric compounds can be natural or modified natural products such as gelatin, starch or carbohydrate derivatives such as methyl cellulose, hydroxyalkyl cellulose, or salts of carboxymethyl cellulose. Polymer of synthetic nature such as poly(vinyl alcohol), partially hydrolyzed poly(vinyl acetate), sodium salts of poly(acrylic acids), methacrylic acids, and copolymers thereof are widely used in quantities as suspending agents.

9.1.6.2 Inorganic compounds

Inorganic compounds are compounds such as alkaline earth carbonates, sulphates, phosphates, aluminum hydroxides, and various silicates used in powder form as suspending agents. They are used in quantities between 0,001 and 1%.

A large number of protective colloids of carefully controlled solubility and molecular weight as well as inorganic solids of controlled particle size and surface wetting characteristics are used. The agitation interacts strongly with the suspending agent.

9.1.7 Inhibitors

Inhibitors can be present as reaction ingredients, which deliberately add to monomers to prevent premature polymerization. They are very active before any significant reaction takes place. Therefore, all inhibitors in feed must be destroyed. Therefore, monomers are present in the reaction to avoid inhibition reaction. Inhibitors are cleaned and handled carefully. Commercial manufacturing processes are usually in operation with inhibitors, which are present in streams. Dead time is commonly the time to stop the reaction, in polymerization reaction, the time is in observation. Inhibitors are in use with batch reaction process before the reaction starts. The most efficient inhibitors are quinones, hindered phenols, and amines [32]. These inhibitors require traces of oxygen to function. Tertiary butyl catechol is the most common inhibitor used in commercial styrene. Nitrophenol, hydroxylamine, and nitrogen oxide compounds [33] are also useful. The removal of inhibitors is essential before polymerization in order to avoid an induction period. It is essential in shipping and storage of monomer such as styrene [34] and sulfur compounds. The retardation effects in polymerization of styrene are due to the presence of metal traces such as iron or copper and sulphur compounds [35, 36].

9.2 Polymerization Processes

Increasing demands require on product quality, process control, safety, and development of improved polymerization processes [37–45]. Polymerization processes are concerned as a continuous process. In a reactor, the monomer, catalyst, etc., undergo continuous stirring. It has evaporative cooling and downstream monomer separation. The monomer is recycled and the polymer

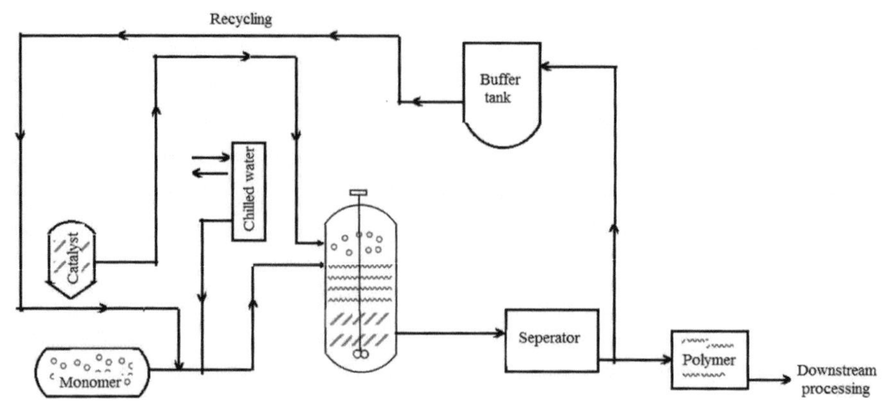

Figure 9.1 Flow diagram of a polymer process.

to process undergoes packing. Either in case of maintenance or sometimes for cleaning, recipe change, modification, or disturbance, the reactor has provision to clear. However, the final polymer melt cannot be buffered [46–48]. The simple flow diagram of a polymer process is given in Figure 9.1. Monomer is either pure or in a solvent enters at one end and leaves as polymer at the other end. Polymerization reactors have usually been concerned with reaction systems in which the viscosity of the solution and problems of heat transfer have not been of direct, concern.

Flow instability, heat transfer, and variation in residence time are important factors to note. In case of the reactor it can no longer be operated, the flow instability occurs due to the plugged polymer or the monomer that comes through largely unreacted polymer due to channeling. Low thermal conductivity of polymer solutions makes heat transfer quite difficult. Side reactions such as crosslinking might occur in remaining material in the reactor causes variation in residence time.

Recipe ingredients are charged and brought to the reaction temperature. In case of initiator it is not part of the original charge, the initiator is added later. The reaction is carried out to the desired degree of conversion. and the polymer is removed for further processing.

9.3 Reactor Types

The reactors classified as batch, semi-continuous or semi-batch, and continuous. A batch reactor is the simplest reactor. In a semi-batch reactor, only part of the charge added is in the beginning of the cycle. From the initial charge, a portion of the monomer is kept aside. A reaction is allowed little time to pass before the addition of the remaining part of the charge in a controlled manner. In a continuous reactor, systems consist of stirred tanks connected with all the recipe ingredients. The polymer is removed at the end, while all the ingredients are fed into the reactor. The use of continuous reactors can behave differently from batch reactors with regard to the polymerization rate.

9.4 Classification

Polymerization processes are important as an evolving branch of technology. A majority of polymers are manufactured by:

- Bulk polymerization
- Solution polymerization
- Suspension polymerization
- Emulsion polymerization

They are the major types of two-phase systems used for the production of polymers. Among these polymerization processes, bulk and solution polymerization utilizes batch or continuous process. The advantages are more uniform products and low volatile levels. These processes result in highly viscous finished products [49–51].

9.4.1 Bulk or Mass Polymerization

Bulk polymerization is very simple (Figure 9.2) and yield polymers with high clarity. Bulk polymerization starts commonly by radical initiators. It is divided into solid and molten phase. A polymer which is obtained by solid phase polymerization is amorphous. It shows no tendency in crystalline. The crystalline matrix is unable to exert any appreciable steric control.

The propagation takes place at the polymer-monomer interface. The local strains and defects in the crystal are controlled by local strains. Polymerization in the molten monomer becomes heterogeneous due to the solubility of a polymer in the monomer itself. The limitation of molecular weights occurs by chain transfer reaction with a monomer. High molecular weight polymers are produced by bulk polymerization [52].

Bulk polymerization consists of heating the monomer without solvent with initiator in a reactor. The monomer and initiator mixture polymerize to a solid shape fixed by the shape of the polymerization vessel. Bulk or mass polymerization is much desirable due to being extremely reactive with

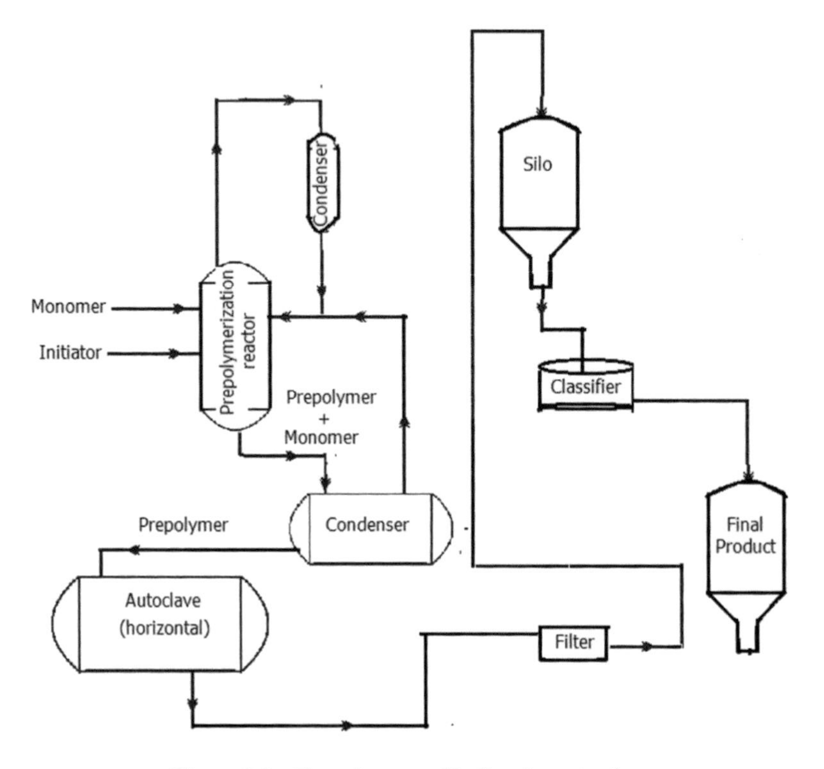

Figure 9.2 Flow diagram of bulk polymerization.

an exothermic Reaction. A higher concentration increases rate of polymerization. Many polymerization processes are carried out in the presence of a solvent. A homogeneous polymerization occurs when both monomers and polymers are soluble in the solvent.

9.4.1.1 Advantages

- Potential for lower cost and economical manufacturing.
- Produces a clean polymer.

9.4.1.2 Disadvantages

- Process needs pure monomer and the presence of impurities affects polymerization.
- Difficulty to remove polymer from a reactor or flask.
- Due to its poor control, this process is not in much use commercially.

- Difficult in the removal of heat from the exothermic polymeric reaction. Removal of reaction heat – Temperature control is difficult due to strongly increasing viscosity [53–55].
- Difficulty in the case of non-solubility of the resulting polymer in the monomer.
- Drastic auto acceleration process may be possible known as gel or Trommsdorff effect. It is necessary to remove heat during polymerization [56–58].
- Side reaction in highly viscous systems such as the Trommsdorff effect (Gel effect) or chain transfer with polymer.
- High conversion, branching, and crosslinking reactions leads to insoluble networks may occur [59–61]. This is due to chain transfer involving abstraction of hydrogen from the polymer chain, subsequent branching and combining two branch radicals.
- Compared with monomers, the volume contraction is high which occurs during polymerization.

9.4.2 Solution Polymerization

Solution polymerization is the most common method and the oldest for production of high molecular weight. It undergoes as batch or a continuous process. In this polymerization, the monomer is diluted with solvents. This dilution makes the temperature control easier. However, solvents reduce molecular weight and polymerization rate.

Solution polymerization is required to ensure a mass flow possible to pump in the reactor. Solvents and residual monomers are removed from the polymer melt at the end of the process. Removal of solvents and residual monomer is realized by heating up the solution followed by a strong pressure drop i.e., flash, thus vaporizing the volatile fraction.

9.4.2.1 Advantages

- Heat removal and viscosity control is easier in solution polymerization.
- Reaction control by choice of solvent of high or low capacity with solvent effects consideration.

9.4.2.2 Disadvantages

- Polymers with poor thermal stability, the overall yield significantly by degrading polymer chain back, mainly into monomer.

9.4.3 Suspension Polymerization

Suspension polymerization is an important mode of production for polymer such as polystyrene, polyvinylchloride, etc. Figure 9.3 illustrates flow diagram of suspension polymerization. The polymerization system contains monomer suspended in water, stabilizing agents, and initiators to speed polymerization. It means where liquid monomer droplets or gas suspended in an aqueous phase under vigorous stirring. It is also called as pearl or bead polymerization. Suspension polymerization of monomers with water solubility, the process behavior is different from bulk polymerization. Non-polar monomers such as styrene and butyl acetate tend to be highly water insoluble. Polar monomers such as vinyl acetate and methyl acrylate are highly water-soluble. Monomers used in suspension polymerization have considerable different solubility. Each monomer droplet contains a large number of live radicals and behave like a small batch reactor. Surrounding water dissipates the heat generated during the polymerization process. Suspension stabilizers or suspending agent is necessary, to avoid the coalescence of the droplets. Particle size, suspending agents, and agitation do not affect generally the rate of polymerization.

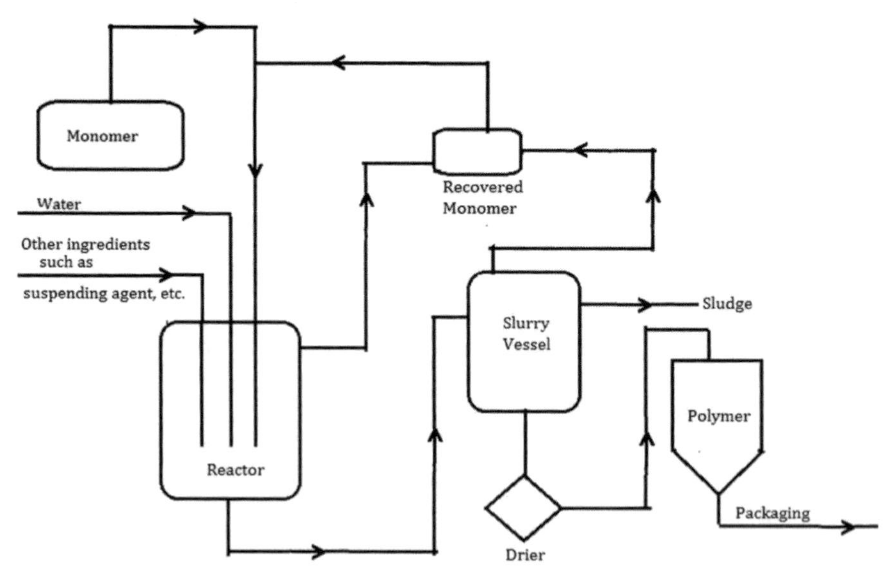

Figure 9.3 Flow diagram of suspension polymerization.

9.4.3.1 Advantages

- Removal of heat and finished polymer.

9.4.3.2 Disadvantages

- Contamination with stabilizing agents.
- Stability of suspension – setup, buildup, particle size and its distribution.
- Problems of optimization are the problems remaining in suspension polymerization.

9.4.4 Emulsion Polymerization

Emulsion polymerization is widely used in industry. Particle nucleation is an important aspect of emulsion polymerization. Micellar, homogeneous and droplet nucleation are three major nucleation mechanism involve in this process. Out of which only one is dominant, depend upon the monomer water solubility and stabilizer concentration [62].

Water is required as a carrier with emulsifying agents. Latex products developed in batch equipment. Monomers emulsified with the help of surface-active agents in water. Free radical initiators added slowly at moderate temperature under an agitation at elevated temperatures which results in the completion of radical reaction with a formation of polymer dispersion. The exothermic polymerization starts and yields extremely small particles. However, the process must be able to produce latex in order to satisfy application requirements at high rate without any frequent disruptions.

Emulsion polymerization is simple process with complex mechanism. The high variability of the polymerization process gives rise to very special conditions. The source of monomers is the emulsified droplets in water. Polymer particles and micelles swollen by monomers are the place for polymerization. Water-phase radicals are polymerized, aggregate or absorb at interfaces. The process has different feeding steps or stirring conditions, temperature profile, etc. [63–67].

9.4.4.1 Advantages

- Emulsion polymerization has ability to increase both rate of polymerization and the molecular weight simultaneously. This ability is not possible in bulk or solution polymerization.
- Due to low viscosity of the mixture, heat transfer during the reaction controls better due to low viscosity of the reaction mixture.
- Health and environment impact is lower in comparison due to absence of large amounts of organic solvents.

9.4.4.2 Disadvantages

- Polymerizations involve free radicals, and therefore control of molecular weight and functionality during the polymerization is difficult to achieve.
- Contamination of polymer with the emulsifier, water, as deficit its clarity and the formation to batch processing.

9.5 Polymer Manufacturing Processes

Three different manufacturing processes are possible, namely,

- Slurry process,
- Gas phase process, and
- Solvent process

9.5.1 Slurry Process

In slurry process, solvents such as isobutylene, hexane, n-alkane are used. Production depends on the variation in pressure, time and temperature. The production of polymers obtained in fine particles form in the diluent. The polymer separation can be by filtration. Chain transfer agents control molecular weight [68–70]. The variation in catalyst design or with varying conditions in several step controls the molecular weight distribution. Long and stirred loop reactors provide best result. With cascade reactors in some processes, allow the variation of chain transfer agents though the operating steps in order to control the molecular weight distribution. The diluent after centrifugation and recycled without purification recovered [71–76].

9.5.2 Gas Phase Process

In addition to the polymerization in bulk or mass, the gas phase process provides polymerization performance without the use of solvents. Therefore, polymerization in the gas phase has advantage that there is no diluent is involved. The process therefore, becomes simple. With supported catalyst, in a fluidized bed reactor, the monomer stirred. In gas phase polymerization, the reaction medium is monomer in the gaseous state or a mixture of monomer with an inert gas such as nitrogen. In this process, polymer granules disperse in the gaseous medium.

The gas conversion maintained by circulation under controlled temperature. As the viscosity of the heterogeneous polymer-monomer mixture is lower, the production with high molar mass is not a critical issue. However, heat capacities of gases are rather low. Controlled gas flows and the polymer periodically removal from the reactor to avoid the blow out of polymer granules from the reactor. The gas circulated to remove the heat of polymerization and fluidizing the bed in order to avoid overheating of polymer granules [77–79].

9.5.3 Solvent Process

The solvent process produces low molecular weight polymer. Solvents like cyclohexane or other appropriate solvents base. Catalyst addition induces rapid polymerization. Water indirectly cools the reactor vessel. The pressure with variation achieves the temperature control [80, 81].

9.6 Summary

- Purification or isolation of the polymers will have residues in bulk material.
- Temperature and pressure control is essential to avoid potential side reactions.
- Incorporation of additives may enhance the polymer processing ability.
- Polymerization under various conditions of temperature and pressure a wide range of polymers made.
- Product from the reactor consists of several homologs in varying concentrations irrespective monomers.
- The final product from conventional reactors consists of an equilibrium mixture of polymer, unreacted monomer, and other products.
- The process variables that affect the process reactivity are the reaction temperature and the concentrations of monomer and initiator.
- Polymerization process consists of five steps: (a) reactor loading; (b) reaction; (c) recovering the non-reacted monomer; (d) treatment of the sludge; (e) centrifuging and drying.

References

[1] M. P. Stevens. *Polymer Chemistry: An Introduction.* Oxford University Press, New York (1990).
[2] G. Odian. *Principles of Polymerization.* Wiely, New York (2004).

[3] J. C. Masson. *Polymer Handbook* (Brandrup, J., and Immergut, E. H., eds), Wiley, New York, II–1 (1989).
[4] C. S. Sheppard. *Encyclopedia of Polymer Science and Engineering*, 2nd Edn (Mark, H. F., Bikales, N. M., Overberger, C. G., and Menges, G., eds.), Wiley, New York, **2**, 143 (1985).
[5] C. S. Sheppard. *Encyclopedia of Polymer Science and Engineering*, 2nd Edn (Mark, H. F., Bikales, N. M., Overberger, C. G., and Menges, G. eds.), Wiley, New York, **I**, 1 (1988).
[6] V. R. Kamath. *Mod. Plast.* **58**, 106 (1981).
[7] R. H. Wiley, N. T. Lipscomb, F. J. Johnston, and J. E. Guillet. *J. Polym. Sci.* **57**, 867 (1962).
[8] S. Machi, M. Hagiwara, M. Gotoda, and T. Kagiya. *J. Polym. Sci. Part A*, **3**, 2931 (1965).
[9] S. Munari, and S. Russa. *J. Polym. Sci.* **4**, 773 (1966).
[10] M. Buback, and H.-P. Vögele. *Makromol. Chem. Rapid Commun.* **6**, 481 (1985).
[11] G. Luft, H. Bitsch, and H. Seidl. *J. Macromol. Sci. Chem.* **11**, 1089 (1977).
[12] P. W. Tidwell, and G. A. Mortimer. *J. Polym. Sci. Part A-1* **8**, 1549 (1970).
[13] Y. B. Malyukova, S. V. Naumova, I. A. Gristkova, A. N. Bondarev, and V. P. Zubov. *Polym. Sci.* **33**, 1361 (1991).
[14] S. Lam, A. C. Hellgren, M. Sjöberg, K. Holmberg, H. A. S. Schoonbrood, M. J. Unzué, J. M. Asua, K. Tauer, D. C. Sherrington, and A. Montoya-Goñi. *J. Appl. Polym. Sci.* **66**, 187 (1997).
[15] J. P. Hogan, and R. L. Banks. *US Pat. 2825721 to Phillips Petroleum Co.*, C.A., **52**: 8621 H (1958).
[16] A. Clark. *Ind. Eng. Chem.* **59**, 29 (1967).
[17] J. P. Hogan. *J. Polym. Sci. Part A-1*, **8**, 2637 (1970).
[18] L. M. Baker, and W. L. Carrick. *J. Org. Chem.* **35**, 774 (1970).
[19] H. H. Brintzinger, D. Fischer, R. Mülhaupt, B. Rieger, and R. M. Waymouth, *Angew. Chem. Int. Ed. Engl.* **34**, 1143 (1995).
[20] P. C. Möhring, N. J. J. Coville. *Organomet. Chem.* **479**, 1 (1994).
[21] L. Resconi, L. Cavallo, A. Fait, and F. Piemontesi, *Chem. Rev.* **100**, 1253 (2000).
[22] G. W. Coates. *Chem. Rev.* **100**, 1223 (2000).
[23] S. Harder, F. Feil, and K. Knoll, Angew. Chem., Int. Ed. 2001, 40, 4261.
[24] K. C. Hultzsch, P. Voth, K. Beckerle, T. P. Spaniol, and J. Okuda. *Organometallics* **19**, 228 (2000).

[25] S. Harder, and F. Feil. *Organometallics* **21**, 2268 (2002).
[26] O. F. Olaj, H. F. Kauffmann, and J. W. Breitenbach. *Makromol. Chem.* **178**, 2707 (1977).
[27] G. Henrici-Olivé, and S. Olivé. *J. Polym. Sci.* **48**, 329 (1960).
[28] T. P. Lodge, N. A. Rotstein, and S. Prager. *Dynamics of Entangled Polymer Liquids: Do Linear Chains Reptate? In Advances in Chemical Physics* (Prigogine, I., and Rice, S. A., eds), John Wiley and Sons, New York, **LXXIX** (1990).
[29] G. Markert. *Houben Weyl: Methoden der organischen Chemie,* part 2 (Bartl, H., and Falbe, J., eds), Georg Thieme, Stuttgart, **E20**, 1157 (1987).
[30] E. Trommsdorff. *Kunststoff Handbuch*, (Vieweg, R., and Esser, F., eds), Carl Hanser, München, **9**, 33ff (1975).
[31] J. V. Dawkins. *Comprehensive Polymer Science*, (Allen, G., and Bevington, J. C., eds), Pergamon Press, Oxford, **4**, 231 (1989).
[32] R. F. Miller, and M. P. Nicholson. *U.S. Pat. 4,465,881 to Atlantic Richfield Co.; C.A.* **101**, 192661q (1984).
[33] J. M. Watson. *U.S. Pat. 4,086,147 to Cosden Techn. Inc* (1978).
[34] J. M. Watson, and W. J. I. Bracke. *U.S. Pat. 4,396,462 to Cosden Techn. Inc* (1983).
[35] M. H. George. *Vinyl Polymerization* (Ham, G. H., ed.), Marcel Dekker, New York, 186–188 (1967).
[36] J. M. Watson, and W. J. I. Bracke. *U.S. Pat. 4,396,462 to Cosden Techn. Inc* (1983).
[37] D. C. H. Chien, and A. Penlidis. Online sensors for polymerization reactors. *J. Macromol. Sci. Rev. Macromol. Chem. Phys.* **1**, 1–42 (1990).
[38] W. D. Hergeth. "Online characterization methods," In *Polymeric Dispersions: Principles and Applications* (Asua, J. M., ed) Kluwer Academic Publishers, Dordrecht, 267–288 (1997).
[39] F. J. Schork. "Process modeling and control," In *Emulsion Polymerization and Emulsion Polymers* (Lovell, P. A., and El-Aasser, M. S., eds) John Wiley & Sons, Chichester, 327–343 (1997).
[40] O. Kammona and E. G. Chatzi, C. Kaparissides. Recent development in hardware sensors for the online monitoring of polymerization reaction. *J. Macromol. Sci. Rev. Macromol. Chem. Phys.* **C39**, 57–134 (1999).
[41] J. Gao and A. Penlidis. Mathematical modeling and computer simulator/database for emulsion polymerizations. *Prog. Polym. Sci.* **27**, 403–553 (2002).

[42] A. Penlidis, J. F. MacGregor and A. E. Hamielec. Dynamic modeling of emulsion polymerization reactors. *AIChE J.* **31**, 881–889 (1985).

[43] E. Saldivar, O. Araujo, R. Guidici and C. Lopez-Barron. Modeling end experimental studies of emulsion copolymerization systems: I. Experimental results. *J. Appl. Polym. Sci.* **79**, 2360–2379 (2001).

[44] E. Saldivar, O. Araujo, R. Giudici, and C. Guerrero-Sanchez. Modeling and experimental studies of emulsion copolymerization systems. II. Styrenic. *J. Appl. Polym. Sci.* **79**, 2380–2397 (2001).

[45] E. Saldivar, O. Araujo, R. Giudici, C. Guerrero-Sanchez. Modeling and experimental studies of emulsion copolymerization systems. III. Acrylics. *J. Appl. Polym. Sci.* **84**, 1320–1338 (2002).

[46] J. P. Congalidis, and J. R. Richards. *Polym. React Eng*. **6**, 71 (1988).

[47] M. Embirucu, E. L. Lima, and J. C. Pinto. *Polym. Eng. Sci.* **36**, 433 (1996).

[48] G. Dünnebier et al. *J. Proc. Control* (2003); H. Seki et al. *Control Eng. Pract.* **9**, 819 (2001).

[49] B. J. Meister, and M. T. Malanga. *Encyclopedia of Polymer Science and Engineering*, 2nd Edn, (Mark, H. F., Bikales, N. M., Overberger, C. G., and Menges, G., eds), Wiley, New York **16**, 46 (1989).

[50] R. Arshady, and M. H. George. *Polym. Eng. Sci.* **33**:865–876 (1993).

[51] C. E. Schildknecht. *Polymer Processes*. Interscience Publishers, New York (1956).

[52] D. Mardare, and K. Matyjaszewski. *Macromolecules* **27**, 645 (1994).

[53] P. Melacini, L. Patron, A. Moretti, and R. Tedesco. *Ital. 903, 309 to Chatillon, SA* (1972).

[54] P. Melacini, L. Patron, A. Moretti, and R. Tedesco. *Ger. Offen. 2,120,337 to Chatillon, SA* (1971).

[55] P. Melacini, L. Patron, A. Moretti, and R. Tedesco. *Ger. Offen. 2,326,063 to Chatillon, SA* (1973).

[56] C. T. Kautter, B. Kösters, P. Quis, and E. Trommsdorf. *Kunststoff Handbuch* **9** (1975).

[57] E. Trommsdorf, H. Köhle, and P. Lagally. *Makromol. Chem.* **1**, 169 (1948).

[58] C. A. Detrick. *Ind. Eng. Chem. Process Design Dev.* **9**, 191 (1970).

[59] B. B. Kine, and R. W. Novak. *Encyclopedia of Polymer Science and Engineering*, 2nd Edn. (Mark, H. F., Bikales, N. M., Overberger, C. G., and Menges, G., eds), Wiley, New York **1**, 269 (1985).

[60] M. S. Matheson, E. E. Auer, E. B. Bevilacqua, and E. J. Hart. *J. Am. Chem. Soc.* **73**, 5395 (1951).

[61] F. R. Mayo. *J. Am. Chem. Soc.* **65**, 2324 (1943).
[62] M. S. El-Aasser, and E. D. Sudol. *Features of Emulsion Polymerization. In Emulsion Polymerization and Emulsion Polymers*, (Lovell, P. A., and El-Aasser, M. S., Eds), John Wiley & Sons: Chichester 37–58 (1997).
[63] R. D. Atkey. *Emulsion Polymer Technology*. Dekker, New York (1991). *Emulsion Polymerization* Applied Science, New York (1975).
[64] R. G. Gilbert. *Emulsion Polymerization, A Mechanistic Approach*. Academic Press, New York (1995).
[65] P. A. Lovell, and M. S. El-Aasser. *Emulsion Polymerization and Emulsion Polymers*. Wiley, New York (1998).
[66] I. Piirma. *Emulsion Polymerization*. Academic Press, New York (1982).
[67] W. V. Smith, and R. H. Ewart. Kinetics of emulsion polymerization. *J. Chem. Phys.* **16**, 592–599 (1948).
[68] W. H. Ray, and R. L. Laurence. *Chemical Reactor Theory* (Lapidus, L., and Amudson, N. R., eds) Prentice Hall, Englewood Cliffs, NJ (1977).
[69] K.-H. Reichert, and H. Geisler. *Polymer Reaction Engineering*. Hanser Verlag, Munich (1983).
[70] D. H. Sebastian, and J. A. Biesenberger. *Polymerization Engineering*. Wiley, New York (1983).
[71] K. Soga, and M. Terano. eds *Catalyst Design for Tailor-made Polyolefins*. Kodansha-Elsevier, Tokyo (1994).
[72] L. Böhm. *Angew. Makromol. Chem.* **89**, 1 (1980).
[73] J. Stevens. *Hydrocarbon Processing* **4**, 179 (1970).
[74] L. L. Böhm. *Catalytic Polymerization of Olefins* (Keii, T., and Soga, K., eds) Kodansha Ltd., Tokyo (1986).
[75] J. Scheirs, L. L. Böhm, J. C. Boot, and P. S. Leevers. *Trends Polym. Sci.* **4**, 408–415 (1996).
[76] L. L. Böhm, and Passing. *Makromol. Chem.* **177**, 1097 (1976).
[77] D. M. Rasmussen. *Chem. Eng.* **79/21**, 104 (1972).
[78] K. Wisseroth. *US Pat. 4012573 to BASF A.-G., C.A.* **77**, 49149 (1977).
[79] W. H. Ray. *Chapter 5 in ACS Symposium Series* **226** (1983).
[80] A. Clark, and G. C. Bailey. *J. Catal.* **2**, 241 (1963).
[81] S. D. De Bree. *Hydrocarbon Process.* **53**, 115 (1974).

10

Future Trends

Polymer chemistry is demonstrating its vast importance in all applications. The developments would be continuing in the future. Polymer overall around and their common structural feature is due to the presence of long covalently bonded chains of atoms. They are versatile and extraordinary class of materials. They are all made from monomers which are attractive with the evolution of technology from its discovery to its present state. To evaluate the current interest in polymer is due to economic cost and mechanical properties.

The monomers are small chemical molecules with characteristics to involve in polymerization reactions. These are attractive for production of polymer with evolution of technology from its discovery to its current state.

10.1 Advances in Polymer Chemistry

Industrial and laboratory chemistry deals with polymer is actually based upon the wide occurrence of the basic element of the structure, benzene nucleus, in the large number of aromatic hydrocarbons isolated during the processing of coal and petroleum. Majority of polymers are industrial products which are obtained by chemical reactions.

Polymers are man-made and synthesized and manufactured due to their durability and resistance to all forms for special performance characteristics which are achieved through control of its molecular weight and functionality. Advances in polymer chemistry have helped to solve problems associated with the studies of monomer to polymer. It helps in synthesis and for the large scale manufacturing.

Science and technology have responsibility in designing synthesis and industrial approaches that are more sustainable. In polymer commonly research aiming to the optimization of polymerization processes and products with respect to material. Polymerization can be an important key in order to produce efficient synthetic processes. The polymerization free conditions in future is important to avoid the risks.

10.2 Advances in Polymerization

Major advances in polymers have continued in contrast to many other areas of industrial research development. Undoubtedly, this advancement is related to polymerization processes and catalysts, understanding polymerization mechanisms, modifying the products, and improving physical properties of the materials. It is important and exciting to research from new findings. Both scientific and technological aspects of polymerization are progress in catalysis, progress in polymerization, progress in polymer design and progress in mechanism-characterization.

Development has a two-fold challenge: finding polymers and finding what properties they have. However, lab syntheses of new polymers have yields only in the milligram to gram range.

Polymers are used for most composites, electronic devices, biomedical devices, optical devices, and precursors for many new developments in high-tech products. Current applications extend from adhesives, coatings, foams, and packaging materials to textile and industrial fibres, elastomers, and structural polymers.

10.3 Environmental Effect

Scientific interest in polymers has grown dramatically, mainly for applications. Polymers are light weight, high rigidity, transparency and inexpensive is designed to be discarded after a single use. Polymer growth is enhanced by their dominant role in their applications. However, disposal of polymers become the notoriety. Polymers have encountered two huge challenges recently. The gradual reduction of petroleum resource has been leading to continuously increased cost of traditional polymer. Furthermore, the environmental pollution caused by non-degradation of traditional polymeric materials has accumulated and endangered lives. However, Polymer degradation shows a single stage degradation involving complex reactions such as intramolecular exchange reactions to form oligomers, hydrolysis reactions leading to the formation of phenolic end groups and CO_2, decarboxylation of the carbonate groups, disproportionation, dehydrogenation, isomerization, etc. [1].

The development of polymer chemistry would be marked by dramatic increases due to its commercial usage in large amounts. However, by the way of manufacturing polymer and to destroy it, a significant proportion of chemical molecules are released in the environment as pollutants. A biodegradable polymer is defined according to standard tests as readily biodegradable if it demonstrates a conversion to CO_2 corresponding to 60% of the theoretical amount during 28 days [2]. Synthetic polymers are in that respect not biodegradable.

10.4 Future Requirements

10.4.1 Pollution Free Polymerization Techniques

The polymer industry is producing and processing huge amounts of polymers. Synthetic or chemically modified polymers belong to a group, it is considered as the polymer enriched group with all the necessary elements in the manufacturing process such as fillers, pigments, antioxidants, processing aids, and flame retarders, etc. They are inexpensive and modifying mechanical properties in a wide range by adding fillers and elastomers. They enhance the comfort and quality of life by keeping hygiene in the modern industrial society. Due to their low density, they occupy high volume fraction despite their relatively low weight fraction [3].

Most polymer systems are inhomogeneous and contain a variety of defects or inhomogeneities at various structural levels such as loops, pendant chains, wide distributions of chain lengths, fluctuations in crosslink or segment density, anisotropic regions and filler particles. Polymer chemistry offers a range of polymers for product development to undergo uniquely useful for many applications.

10.4.2 Polymerization with Solvent-Free Conditions

- Reduces risks inherent to the use of high amounts of organic solvents in polymerization.
- No solvent to no recovery, purification, and reutilization.
- Reducing solvents avoid pollution to environment.
- Easy recovery of the polymers without solvent impurity.
- Oxidation or reduction reactions during processing of polymer with solvent may require more additives.
- Rapid reactions with reactions of monomers.
- More economical for industry, due to lack of solvent.

10.4.3 Waste Reduction

The increasing global environmental awareness, depletion of petroleum resources, drive for sustainable technology and regulations have triggered the search for new products and processes. Examples of renewable raw materials which are used industrially are cellulose, starch, sugar, oils and fats. The philosophy of the renewable management of raw materials is that only amounts which do not upset the natural biological equilibrium are extracted, and thus no disruption of the biosphere results.

The use of polymers in disposable consumer goods has grown tremendously. This growth is proving on the consuming a large fraction of available landfill space in waste disposal system. Synthetic polymers are obtained from petroleum, as a limited, non-renewable resource. Polymer wastes are unreasonable wastes of high quality materials. It contributes the growing problem of the global waste disposal. In domestic sources, contaminated mixed polymers present. It is a difficult challenge to traditional recycling techniques. Hydrocarbon polymers do not biodegrade rapidly in compost or in soil.

Highest priority is to reuse and recycle polymers to handle after their projected service time. Recovery of polymers includes mechanical and chemical feedstock recycling as well as energy recuperation. Burning of polymer is the last priority of recycling and can be used to save natural resources. Reprocessing is also recycling and the polymer waste can be used directly in plant. Waste deposit limitation being the most important [4, 5].

The use of biodegradable polymers to reduce the environmental pollution caused by polymer wastes. There is an urgent need to develop renewable resource-based, environmentally benign materials, which will reduce the problem of plastic wastes and will be a solution to the uncertainty of the petroleum supply.

10.4.4 Polymer and Human Beings

The properties of simple molecules are independent of their physical history. Small molecules are completely determined by the nature of molecules. Polymers however, are not capable of adjusting themselves instantly to any changes in physical environment. The properties of polymer may vary over a wide range depending upon the physical treatment it has received.

The key considerations are that biocompatible polymer does not cause any adverse reactions when in contact with human cells. Moreover, they provide building blocks for the reconstruction of extracellular matrix components. On the other hand, biocompatible polymer such as chitosan is recognized by tumour cells, and therefore, it can bring drugs to their target selectively. Therefore, medical and pharmaceutical applications can easily be worked out with joint efforts from polymers with various fields. Polymer intends to provide interdisciplinary insight in the scientific knowledge immediately usable to realize the potential in the pharmaceutical field.

10.5 Green Polymer Chemistry

The implementation of green polymer chemistry in the field of polymer science includes design and synthesis of sustainable polymers. Such polymers should be produced from feed stocks derived from renewable or biomass resources by environmentally benign processes, such as enzymatic and solvent free processes. It avoids the use of hazardous materials. However, there should be possibility of recycling and biodegradable. Bio-degradable polymers make a realistic contribution to the recovery of value from waste packaging as fertilizers and soil improves for agriculture and horticulture.

The high performance material leads to a reduction in its consumption is very important. A chemically recyclable and biodegradable polymer should contain enzymatically cleavable linkages such as ester and carbonate linkages. Therefore such polymer chain can be broken by environmental microbes during biodegradation. In chemical recycling, polymer can be cleaved by specific enzyme into oligomers or monomers that can be re-polymerized in the reverse reaction of the enzyme.

Research covers a wide range of specifications, e.g., the development of new materials synthesized and classic organic chemistry and improvement of material properties, including degradability pathways and patterns [6–8] as well as mechanical and physical properties, with much interest in the investigation of biodegradability itself, searching for measuring techniques and assays [9].

10.6 Summary

- Polymer is one of the most important materials in which multidisciplinary applications contributing to engineering integrated technology, consumer applications and to human health care.

- Polymerization is a viable process for the production of polymer from monomer.
- Polymer synthesis run on a large scale which pose disposal problem which is very costly.
- Recovery of useful materials from polymer waste depends on number of factors such as ease of recovery, cost of recovery, degree of purity, potency of recovered materials, and most important whether materials are recoverable or not.
- Green polymers are polymers derived from renewable resources.

References

[1] G. Sivalingam and G. Madras. *Ind. Eng. Chem. Res.*, **43**, 7716–7722 (2004).

[2] OECD *Guidelines for Testing of Chemicals*. OECD, Paris (1981).

[3] M. Thayer. *Chem. Eng. News*, April 5, 7 (1989).

[4] R. Narayan and S. Blomembergen. *J. Am. Chem. Soc. Polym.*, Preprints **32**, 119–120 (1991).

[5] M. M. Nir, J. Miltz, and A. Ram. *Plastics Eng.*, **3.93**, 75–93 (1993).

[6] D. L. Kaplan, J. Mayer, S. Lombardi, B. Wiley, and S. Arcidiacono. *J. Am. Chem. Soc. Polym.*, Preprints **30**, 509–510 (1989).

[7] J. E. Glass and G. Swift. *Agricultural and Synthetic Polymers-Biodegradability and Utilization, ACS Symposium Series 433*. Washington, DC (1990).

[8] K. Sakai, N. Hamada, and Y. Watanabe. *Agr. Biol. Chem.*, **50**, 989–996 (1986).

[9] J. Augusta, R.-J. Müllller, and H. Widdecke. *Chem. Ing. Tech.* **64**, 410–415 (1992).

Index

About the Author

Muralisrinivasan Natamai Subramanian is a plastic technology consultant specializing in materials, additives, and processing equipment, including troubleshooting. The author obtained his B.Sc in Chemistry from the Madurai Kamaraj University and his M.Sc (1988) in Polymer Technology from Bharathiar University. He received his Post Graduate Diploma in Plastics Processing Technology from CIPET, Chennai. He has also completed his Doctors of Philosophy in Polymer Science from Madurai Kamaraj University. He worked in the plastic process industry, mainly in R&D, for 13 years before turning to consultancy and building up an international client base. Muralisrinivasan conducts plastic processing seminars as well as serves as a Board of Studies Expert member of Colleges in India dealing with curriculum of technology subjects. He authored *Update on troubleshooting in Thermoforming* in 2010, *Update on troubleshooting PVC processing* (iSmithersRapra), *The basics of troubleshooting in plastics processing* (Scrivener-Wiley Jointly), *Update on troubleshooting in PVC Extrusion process* (iSmithersRapra) and *Plastics testing – New instrumental methods* (Momentum press).